Herman Biddell

Heavy horses

breeds and management

Herman Biddell

Heavy horses
breeds and management

ISBN/EAN: 9783741196515

Manufactured in Europe, USA, Canada, Australia, Japa

Cover: Foto ©Klaus-Uwe Gerhardt /pixelio.de

Manufactured and distributed by brebook publishing software (www.brebook.com)

Herman Biddell

Heavy horses

LIVE STOCK HANDBOOKS.

Edited by JAMES SINCLAIR, *Editor of* "*Live Stock Journal*," "*Agricultural Gazette*," *&c.*

No. III.

HEAVY HORSES.

BREEDS AND MANAGEMENT

BY

HERMAN BIDDELL; C. I. DOUGLAS;
THOMAS DYKES; DR. GEORGE FLEMING C.B., F.R.C.V.S.;
ARCHIBALD MACNEILAGE; GILBERT MURRAY;
AND W. R. TROTTER.

THIRD EDITION.

ILLUSTRATED.

London:
VINTON & COMPANY, Ltd.,
9, NEW BRIDGE STREET, LUDGATE CIRCUS, E.C.

1898.

CONTENTS.

	PAGE
CHAPTER I.—The Shire Horse	1
CHAPTER II.—The Suffolk Horse	37
CHAPTER III.—The Clydesdale Horse	75
CHAPTER IV.—The Breeding of Heavy Cart Horses for Street Work ...	122
CHAPTER V.—The London Work Horse in Street and Stable ...	145
CHAPTER VI.—Farm Management of the Heavy Horse	159
CHAPTER VII.—Diseases and Injuries to which Heavy Horses are Liable	183

ILLUSTRATIONS.

	PAGE
Cart Horse, Dodman ... ...*To Face*	9
Shire Stallion, Harold ... . ,,	13
Shire Mare, Blue Ruin ... ... ,,	14
Shire Stallion, Bury Victor Chief ... ... ,,	17
Shire Mare, Rokeby Fuchsia ... ... ,,	18
Shire Stallion, Vulcan... ... ... ,,	21
Shire Mare, Lockington Beauty ... ... ,,	22
Shire Mare, Starlight ... ... ,,	24
Shire Stallion, Dunsmore Willington Bay ... ... ,,	26
Shire Stallion, Bar None ... ... ,,	28
Shire Stallion, Staunton Hero ... ... ,,	31
Shire Stallion, Markeaton Royal Harold ... ... ,,	33
Shire Stallion, Rokeby Harold ... ... ,,	35
Shire Stallion, Prince William ... ... ,,	36
Suffolk Mare, Bounce... ... ... ,,	54
Suffolk Mare, Bramford Belle ... ... ,,	56
Suffolk Stallion, Prince Wedgewood ... ... ,,	64
Suffolk Stallion, Eclipse ... ... ,,	70
Suffolk Mare, Queen of Hearts ... ... ,,	72
Clydesdale Stallion, Flashwood ... ... ,,	103
Clydesdale Stallion, Macfarlane ... ... ,,	104
Clydesdale Stallion, Prince of Carruchan ... ... ,,	106
Clydesdale Stallion, Top Knot ... ... ,,	115
Clydesdale Mare, Laura Lee ... ... ,,	116
Clydesdale Stallion, Laurence Chief ... ... ,,	118
Clydesdale Mare, Sunrise ... ... ,,	121
Dray Horses... ... ... ,,	146
Shire Geldings ... ... ,,	157
English Dray Horses ... ... ,,	180

HEAVY HORSES.
BREEDS AND MANAGEMENT.

CHAPTER I.
THE SHIRE HORSE.

THE task of attempting to give a short history of the rise and progress of the Shire horse is by no means an easy one. It is true that a considerable amount of evidence exists to show that in days long before the Christian era the breed of horses in these islands was considered to be, when viewed from a Roman standpoint, unusually large; but it should be remembered that size is always more or less comparative, and as we have no plates or drawings to guide us when speaking of the very early horses, we can only surmise that they were really of very considerable bulk, because their descendants were decidedly large, and bore a very close resemblance to the Shire horse of the present day. It must, however, be recollected that the horse of a certain period is naturally moulded so as to be suitable to the requirements of that time; and the Great Britain of two thousand years ago was, of course, a vastly different country from the Great Britain as we now find it. The nation then was troubled and disturbed, and the majority of the inhabitants may be described as warriors. This state of things created a demand for horses fitted to carry armed men, weighing, together with their armour, probably about 400 lbs., a fact that in itself proves that size and weight-carrying capacity must have been present in the horses of that day. We also have to bear in mind that the surface of the country at that time was rough, wooded and rugged, and therefore a fast light horse would not have been so much in request as one that possessed strength and substance.

In these suppositions we are confirmed by the written evidence that exists, and by the other records of the times, for Cæsar himself recounts the methods of warfare carried out in Great Britain in those days, and mentions the chariots full of warriors that were drawn at a rapid rate over the rough and uneven ground, thus demonstrating that the horses that drew them must have possessed weight, substance, and activity. In dealing with this portion of the early foundation stock, we are sure we cannot do better than refer to Sir Walter Gilbey's eminently useful and interesting little work on "The Old English War Horse," known in these latter days as "The Shire Horse." As far as the actual antiquity of the breed is concerned, all readers of that book must certainly have come to the conclusion that Sir Walter Gilbey very clearly proves the existence in Great Britain of horses of unusual weight and size at the time of the Roman Invasion, or, in other words, considerably over 2,000 years ago.

It is not necessary now to closely follow the arguments that are adduced in "The Old English War Horse" in order to convince readers of to-day of what may well be taken for a fact, namely, that large weighty horses existed in very early periods; but of course it is not contended that the Shire horse, as we find him now, prevailed at that period. Putting, however, this original early foundation stock on one side for the present, we propose in the first place to pass over a space of some 1,600 years, which will bring us down to the year 1505 A.D., subsequently commenting slightly on the intervening period. In Sir Walter Gilbey's book the earliest plate to which we are introduced is one by Albert Durer, dated 1505. It represents a horse of that period with a soldier in armour standing beside it, and the first thing that strikes one is that either "there were giants in those days," or that the drawing is out of proportion, because the man towers considerably over the horse, while the conformation of that animal would lead us to suppose it to represent a horse of very large size, weight and substance. These, however, are only minor details, whereas what we really have to study is the character and class of horse represented. With regard to these points there can be no doubt or misunderstanding, for we find plainly placed before us what may well be termed a "modified Shire." The profuse hair in mane, tail and fetlocks is present; the wide hips and slightly drooping quarters, the large spreading feet, the massive shoulders and swelling muscle, all denote a horse of great size; still, one can detect from the class of bone and the sloping pasterns that his duties were by no means confined to slow work, but that the exigencies of war required of him a faster gait. With these few exceptions, however, we see that the horse before us, painted in the year 1505, s very similar to what we now denominate a "Shire," but we must take into consideration the purpose for which the animal was then used, the probable hardships he had to undergo, and the want of that care and high-feeding which have latterly become part and parcel of his existence.

It is not to be supposed, because we place the year 1505 as the point of departure, that previous to that time heavy horses did not exist; indeed, as we have before pointed out, written evidence proves the contrary, but we select the year named merely because the picture in question proves at one glance what we require, and gives us a starting point from which we can work backward and analyse the early evidence, and forward through the pictures and history of those times. Reverting, then, once again to older periods, we find that several very interesting extracts occur as far back as the twelfth century, and in the time of Henry II. we also observe that an effort was made to improve the horses of this island, for in the year 1160 we read of the importation of breeding animals from abroad, and the probability is that they were of a heavy class, as in 1154 William Stephanides speaks of a fair at Smithfield where "Cart horses fit for the dray or the plough" were exposed for sale. A few years later, in the reign of King John, we read of a further importation of a hundred stallions of large stature from Flanders, Holland and the banks of the Elbe, probably also of the same variety.

After this period up to the time of Henry VIII. the record of progress is hardly so clear and defined; yet it is worthy of note that various Acts of Parliament were passed, all intended to increase the size and weight of the then existing horse. The picture by Durer, to which we have alluded, was painted prior to the accession of Henry VIII., and although, as we have seen, it shows all the characteristics of the Shire, still, it would seem that the required uniformity of size, height and weight had not been sufficiently attained in Henry VIII.'s time, as very stringent laws were then enacted against the running at large of inferior animals, and further it was prescribed that if within certain counties any such specimens were found they were to be destroyed. Another somewhat curious Act was also put in force in that reign, by which it became a felony to export any horses. Scotland was one of the countries into which no horses were allowed to be taken; and this latter clause is rather interesting to-day when the antiquity of the Shire and Clydesdale breeds is compared. About this period we find an extract from a book written by Sir Thomas Blundeville, and the following quotation occurs in Sir Walter Gilbey's work :—" Some men have a breed of great horses, meete for the war and to serve in the field ; others tried ambling horses of a meane stature for to journey and travel by the way. Some, again, have a race of swift runners to run for wagers or to gallop the bucke, but plane countrymen have a breed only for drafts or burden." Now this passage, when analysed, affords very strong evidence that at that time (now nearly 500 years ago) the different breeds existed much as we find them to-day. They evidently had the Race horse, the Hackney, and the Draught, or Shire horse as he is now called.

These, then, are points that are well worthy of notice, and they furnish a very distinct contradiction of the expressed opinions of certain individuals that there is only one pure breed in England, and only one worthy of a Stud Book, and that that breed is the Race horse. This contention is clearly ridiculous, because, as far as England is concerned, everything decidedly points to the original horse having been very far removed from the thoroughbred, and the efforts towards improvement carried on for centuries were all in the direction of increasing size and weight, long before attempts were made to introduce the Arabian.

Coming still nearer to the present day, it is curious to note the different values of the several classes of horses in the year 1620, at which time an estimate was laid before the Privy Council for horsing a foreign expedition. From this it would appear that the ordinary class of animal was quoted at £9, while the "strong, or great horses" were estimated at £15, again proving them to be valued in those days at nearly double the price of the other varieties. And if we now compare the ordinary animal of our own time with the best type of Shire, we find that the comparative values are much the same, putting, of course, pedigree to one side, and only calculating the market value of work horses. The point of similarity between the old type and the present is also found in an extract from the work by the Duke of Newcastle, published in 1658, profusely illustrated with plates of the Great Horse of that time, and among other excellencies, drawing attention to action. Now, the Shires at present are especially strong in this point; in fact, none of our breeds show the same large proportion of action, both in the walk and trot, as does the Shire. This, of course, many will be inclined to dispute, but nevertheless, it can be very easily proved at any time. Those interested in this subject can amuse themselves by noticing the horses at work in any large town, and they will soon be convinced that the heavy breeds, the descendants of the "Great Horse," show their shoes fully every step they take in a very different form to the lighter varieties.

In 1713, Queen Anne was an enthusiastic admirer of the Shire, as her state equipages were horsed by long-tailed mares of this breed; but about that period it would seem that some of the ancient methods of locomotion had been more or less abandoned, and the saddle was giving place more and more to the coach; probably, also, the roads were in better condition than before, and consequently, a new breed was sought to be established by crossing the Shire mare with the blood horse, and thus creating the Coach horse.

In 1796, or about one hundred years ago, the *Sporting Magazine*, after nine years' existence, speaks of the Draught horse as being equal to moving a load of three tons singly, and if we allow for inferior roads, we must admit the Shire of that time to have been equal in moving power to the Shire of to-day.

EARLY STUD BOOK RECORDS.

Having so far attempted in a slight measure to weld together the past with the present, and to trace the fortunes of the original war horse through several centuries, until civilisation has converted him into one of the principal factors and aids in our modern commercial life, let us now once more turn back, for we are getting beyond the region of conjecture and entering upon a period when we have to deal with indisputable facts, to a time when records begin, when we can give day and date, chapter and verse; in fact, we shall now have to turn to Volume i. of the Shire Horse Stud Book. Before doing so, however, we wish once again to point out that in dealing with olden times and foundation stock, the account we have attempted to give has necessarily only been a sketch, because the evidence existing in very early times is, of course, only circumstantial and inferential; still we think it will generally be conceded that all the history we have on the subject of the early horse points towards size as being especially desired, and that as far as antiquity is concerned, there can be no doubt but that the foundation stock that originated the Shire was to be found in England over 2,000 years ago. Those who have thus far followed us will no doubt feel relieved that we have now left behind the days of war horses and ancient English, and have emerged into the more congenial atmosphere of modern requirements, when we have to deal with correct joints, silky feather, hocks well together, and the other various requisites of the Shire horse of our time.

Before this is quite possible, we must for a moment glance at the early Stud Book days, when in fact the work of dividing the wheat from the chaff really began. It cannot have been a light labour, and certainly all honour is due to those who first initiated the idea of a Stud Book, and to those who devoted their time and brains to turning chaos into order. No doubt there are many people in some measure interested in our great breed of Shire horses, who do not devote much time to studying the early volumes, but rather interest themselves in those books in which their names, and the names of the animals they have bred, appear. Consequently these have, generally speaking, but little knowledge of the early entries, and therefore, as we still wish to link the chain of evidence together as closely as possible down to the present day, we desire to put in print here the information about some of our recorded foundation stock, a very essential and necessary thing to do, as it appears to us, especially as it seems that even now there are those who are so ignorant as to attempt to throw doubt on the antiquity and purity of this breed of heavy horses.

In commenting on the first volume, it will be necessary at

the outset to review and repeat portions of the introduction of that work, which contains the names and other particulars only of animals foaled previous to the year 1877. Mr. Reynolds's instructive "History of the English Cart Horse" in the Stud Book, has doubtless been studied by members of the Shire Horse Society, but it may not be out of place here to again draw attention to some of the leading points, because having now been written fourteen years, Mr. Reynolds's views will well stand repetition, if only as a warning to some breeders, who to-day would endeavour to entirely depart from the old traditions and characteristics of the breed, and create a new Shire horse moulded to suit their own special ideas. To-day one can hardly properly appreciate the amount of labour it must have taken to collect the names and breeding of the 2,381 stallions whose pedigrees we find in the pages of the first volume of the Shire Horse Stud Book, the compilation of which was made particularly intricate owing to the fact that every animal was born previous to the year 1877, and that it went back to the year 1770, thus covering a period of no less than 107 years.

Before taking notice of a few illustrious and well-known names that we find within this period, it may be desirable to review some particulars of the breed in the different counties as furnished in Mr. Reynolds's carefully prepared article. It is noticeable that he draws attention to what of course is an undisputed fact, viz., the admixture of the foreign element in the composition of the Shire. This no doubt is so; in fact we have already pointed out, while referring to Sir Walter Gilbey's book, some of the periods at which importations of stallions from the Continent of Europe occurred, and we must say it seems rather hard to accuse the Shire horse of impurity because of the efforts on the part of our forefathers to increase the standard of size by the introduction of foreign stallions, particularly when we recollect that nearly all breeds of domesticated animals, as we now find them, are more or less the creation of the breeders.

CART HORSE, DODMAN (Foaled about 1780).
Owned by an Ancestor of Mr. Anthony Hamond, Westacre, Norfolk.
(From Sir Walter Gilbey's Book, "The Old English War Horse or Shire Horse.")

having been brought into existence through the requirements of the times and the advance of civilisation. On the other hand, it cannot be denied that evidence still exists, and no doubt years ago existed in a still greater degree, to prove the prevalence of at least two varieties of the Shire horse. We refer to those endowed with peculiar hirsute appendages, such as the moustache on the upper lip, and the long lock of hair hanging from the knee, and also projecting from the back of the hock. In this variety the hair is also found in profusion hanging from the back of every leg. It is also true that in certain strains we find an absence of the first three peculiarities, and a general lessening of the quantity of hair in other parts. This latter strain would certainly appear to have an infusion of light blood of some description, but whether it was derived from foreign ancestry is open to doubt. Mr. Reynolds records that the moustache was in older times considered a peculiarity of the Lincolnshire-bred animal, and as we have pointed out that the tendency to excessive growth of hair accompanies this feature, we may conclude that for a very long period indeed, the Shire horse of Lincolnshire and the East Coast has been noted for what we now term "sourness." Having, then, endeavoured to localize the "Horse of Hair," let us ask, Whence did the variety probably spring? For one moment we again refer to Sir Walter Gilbey's book, and towards the end we find a plate of the horse Dodman foaled in the year 1780, and the property of an ancestor of Mr. Anthony Hamond, of Westacre, Norfolk. This plate of the horse Dodman shows us the hairlock from the knee growing clearly defined, and hanging nearly to the fetlocks; the hair at the back of each leg being also present in great quantity. Unfortunately the breeding of Dodman has not been traced, as his name does not appear in the pages of the Stud Book, but we think we may take it for granted that he was at least an East Anglian horse, and a specimen of the old variety in that locality.

We now propose, therefore, to account, by probabilities, for the origin of these distinctive horses in that portion of Great Britain, viz., the East Coast. In our opening remarks we pointed out that the histories and traditions of the early centuries are more or less involved in doubt and mystery; still, although we are unable to appeal to historical data, we have a certain groundwork to go upon, and many probabilities upon which theories may well be built. Thus, although we have already mentioned certain years in which importations from abroad undoubtedly took place, it can hardly be questioned that long before those periods many horses of the heavy variety were brought from the Continent. We know that the first recorded importations from Flanders were in the year 1160, and it is not unreasonable to suppose that in those early days transport was surrounded with very considerable danger, and that the very shortest cut from land to land would be taken advantage of. By looking at the map, the first thing that strikes one is that Amsterdam, in Holland, is directly opposite to the Norfolk Coast, and therefore, as we already know that Holland was the country from which stallions were selected, the natural conclusion to arrive at is that they were landed in the neighbourhood of Yarmouth, and thence gradually spread westward, either themselves or through their progeny, and that the development of the best specimens was brought about by the richness and suitability of the soil in the different counties where they made a lodgment. This proposition is only put forward as theory, but nevertheless it is a very probable theory, and one founded on a very reasonable basis. We would not, however, wish it to be supposed that a horse precisely similar to (let us say) Dodman was imported, because the evidence we possess bears quite the other way; still our idea is that, at least, the required size and weight came from abroad, and the accessories were brought about by centuries of crossing on the native animal, by the nature of the soil, and the fashion of that day, which probably tended in the direction of cultivating hair because it was found that hair and substance went together.

Pedigree Influence and Traditions.

The gradual distribution throughout the richer counties no doubt was slow, and must have only kept pace with the improvement of the land and the requirements of the farmers; but there is no doubt upon one point: that very many years ago the shires of Nottingham, Leicester, Derby, and Stafford were celebrated for the excellence of their cart horses, the limestone in those localities, doubtless, having immense influence; in fact, when we turn to the chronological table, given in the first volume of the Stud Book, we see that the honour of providing the oldest pedigreed horse belongs to Leicestershire, the animal being Blaze 183, foaled in 1770. In 1773, there is Ruler 1905, a Derbyshire-bred one; in 1775, G. 890, a Leicestershire horse; and in 1778, Bald Horse 93 also hailing from the same county.

Having thus made a start in the heart of the Midlands, it is not in the least degree surprising to find that the large majority of the stallions that date back farthest in the Stud Book had their birthplace in one of the counties mentioned. Various other names occur incidentally of other horses belonging to an earlier period still; for instance, the Packington Blind Horse, who is credited with having been in the full vigour of his existence some fifteen years prior to the earliest Stud Book record, or, in other words, in the year 1755. Then we have Mansetter (Oldacre's) who was the sire of Blaze 183, and who must have been a contemporary of the Packington Blind Horse. In this connection it is curious to note that the name "Blaze" occurs no less than thirty-five times in succession in the first volume, and that only two out of the whole number of horses that were thus named are described as having been brown in colour, the other thirty-three having been black or grey, two colours Mr. Reynolds draws attention to as being indicative of pure breeding, colours that, unfortunately, in these days are conspicuous only by their absence. It would seem that considerably over

100 years ago stallions earned remunerative fees, for it is on record that a grandson of the Packington Blind Horse travelled at three guineas, and Sweet William was sold in 1778 for 350 guineas; while a few years later Marston fetched 500 guineas. Thus we see that even then a good Shire sire was appreciated, and the breeding was doubtless known full well for many generations back, by the families who kept stallions. The pedigree table in connection with the Packington Blind Horse is interesting, as, although his birth was supposed to have taken place about the year 1760, yet his direct descendants are traced down to the year 1832. At the same time this tracing is only in a measure satisfactory, because it is confined almost, if not entirely, to animals bred within a somewhat small area, namely, a portion of Leicestershire and Derbyshire. This was, of course, unavoidable, as the interest in Shire breeding had then been greater in the Midlands than in Lincolnshire and the East, and, moreover, better records were kept; but it would have been well had it been possible to have gone somewhat further back, and discovered the direction from which these Derbyshire horses originated. Our idea, as we have before stated, is that they all came in the first place from the East, and that the old-fashioned Shire horse was originated there. Here we may state what is an incontrovertible fact, if a strange one, namely, that East Anglia has always been and still continues to be far in advance of West Anglia in the breeding of every variety of the horse, and for some mysterious reason the West never seems to be able to get on even terms with the East.

In looking carefully through certain pedigrees of more recent horses, we find that nearly all those that have left their mark in the Midlands—horses to which breeders trace back with pride—originated in the East, and usually through the dam they find their way back in that direction. Take, for instance, Champion 419 (Styche's). He was bred in Oxfordshire; his sire, Rippendon's Champion, is, unfortunately, not traced, and consequently not numbered, and one is therefore

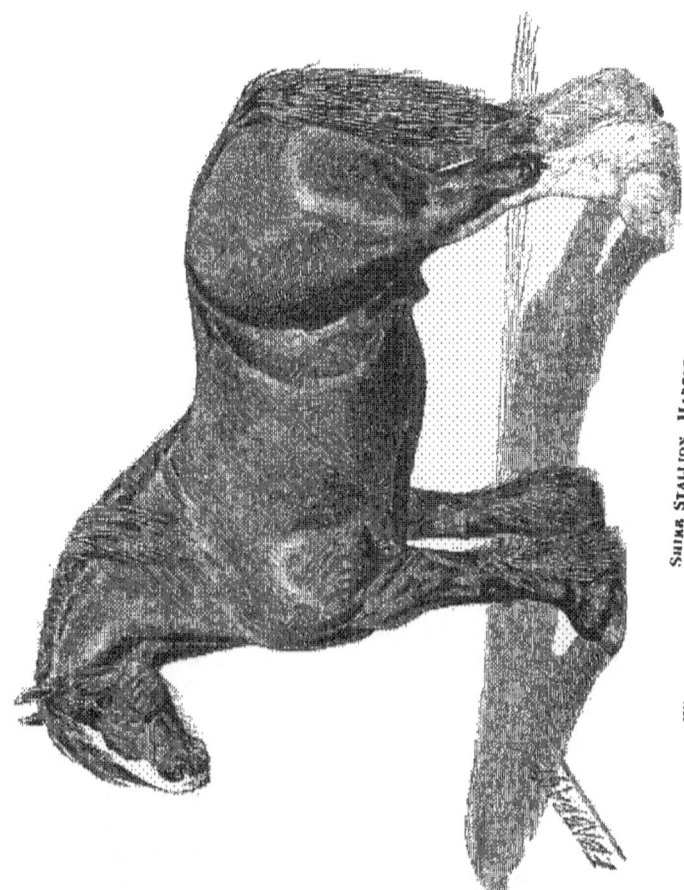

Shire Stallion, Harold.
Winner of Elsenham Challenge Cup, London Shire Horse Show, 1887.
The Property of Mr. A. C. Duncombe.

unable to test his origin; but the dam of Champion 419 was by Conqueror 529, also an Oxfordshire bred one, but by Champion 379, bred in Northamptonshire, got by Farmer's Glory 818, also bred in Northampton. Thus we see the tendency to revert Eastward to Lincolnshire and Cambridgeshire, the adjoining counties; and throughout all these crosses the same grey colour had been retained during all the North-Westerly wanderings. Now, these same Styche's Champion 419 descendants must be considered as quite in the front rank of recent Derbyshire Shires, and it is at least probable that a very considerable proportion of his prepotency arose from his descent from Eastern ancestors, evidently of undiluted origin, this being emphasised by the fact of the grey colour continuing intact through many generations. K., *alias* Lincolnshire Lad, *alias* Honest Tom 1196 is another illustration of the same theory. This horse is commonly known as Drew's Lincolnshire Lad. He, of course, is a direct example of importation from Lincolnshire, for he was bred by Mr. Bassitt, Willoughby, Spilsby, Lincolnshire, and sired by Lister's Lincoln. This horse is unnumbered, but as Mr. Lister lived at Saleby, Lincolnshire, there can be no doubt as regards his thorough Lincolnshire origin on the sire's side, and as the dam was also sired by Mr. Lister's Briton, and ran back to Competitor 514, the property of Mr. Clark, Murrow, Long Sutton, Lincolnshire, we find in following this pedigree that we gradually get nearer and nearer to the sea. Take, again, William the Conqueror 2343, who was by Leicestershire 1317, by Ben 120, and so on to Nelson 1609, tracing to Bald Horse 93, foaled in 1778. On the other side of the pedigree, the evidence is clear, the dam having been by William the Conqueror 2340, by William the Conqueror 2339, by Leicestershire 1321, by Blacklegs 142, by Derbyshire 577, and lastly by Honest Tom 1062, a horse foaled in 1806 at Swarby, Lincolnshire, and, on both sides, of old-fashioned Lincolnshire blood. This old horse was, by the way, a very noted animal in his time, and went under various names, such as Old Tom, *alias* Little David, *alias* Old David, as well as Honest Tom. He evidently was in those early days—some ninety years ago—as celebrated in Lincolnshire as his descendant was about sixty years afterwards in Derbyshire, for we read that he was sold to a Mr. Casswell when he was five years old for 300 guineas, and his service fee was three guineas a mare.

Leaving now the counties of Derbyshire and Leicestershire, let us take up a Nottinghamshire celebrity, What's Wanted 2332, and although his sire was also Nottinghamshire-bred, still we very soon find the same origin cropping out, for his grandsire, Matchless 1609, was a Lincolnshire horse, and his ancestry on the dam's side ran back to Honest Tom 1060, foaled in 1800, and standing next but one in the Stud Book to Honest Tom, *alias* Little David, the ancestor, as above stated, of William the Conqueror.

Before leaving the pedigree catalogue, we are tempted to insert one more name—that of a famous horse that has, unfortunately, lately joined the great majority. We refer to Premier 2646, the most noted son of What's Wanted 2332. In him we have a curious example of wandering parents. Premier, as most people know, was bred in Lancashire, and became famous in Derbyshire, but as his sire was What's Wanted, and his dam also traced to Eastern origin, we find still another case verifying our theory.

It would be impossible, and also tend to make this account very unwieldy, if we were to continue making tables to prove our point; space would not permit of it, and therefore the instances given must suffice. It may be said, however, that the above names have been taken quite at random. The reason they were introduced here was simply because they are, perhaps, as well or better known to most people than any others. We have, however, perfect faith that were it possible to trace the pedigrees to the foundation, every horse in the Stud Book would be found to originate in the East, and the further back the record went the nearer to the German Ocean

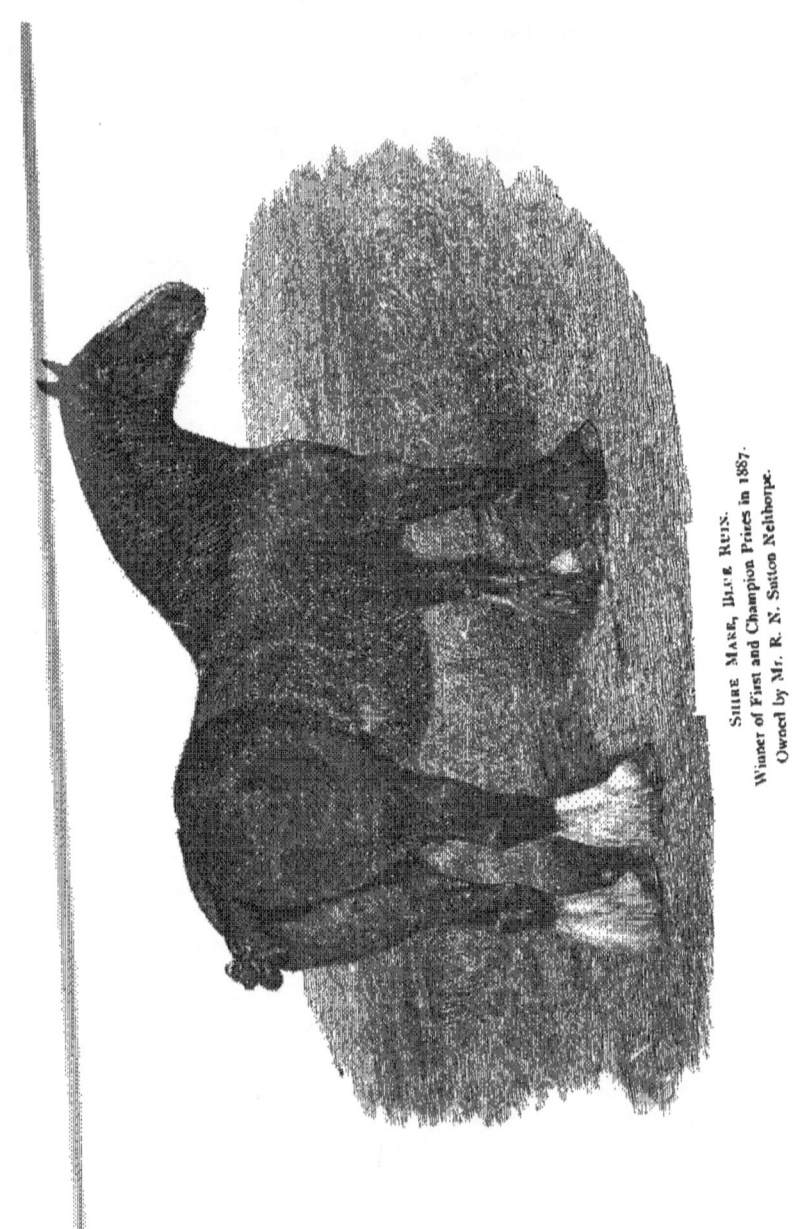

SHIRE MARE, BLUE RUIN.
Winner of First and Champion Prizes in 1887.
Owned by Mr. R. N. Sutton Nelthorpe.

would we get, till eventually the evidence would carry us over the water. This method of investigation would only be found to apply to any great extent to animals that showed in their characteristics a pure origin, or, in other words, that displayed a tendency towards a development of hair in the form and at the points that have been indicated. Mr. Reynolds would appear to believe that the clean-legged varieties of the Shire horse owed this cleanliness to the foreign element, but we incline to doubt that theory, unless he refers to recent foreign blood (which probably he does), and not to out-crossing through the old original importations. If the allusion is to recent importations, his theory is in a measure correct, because we undoubtedly recognise this adverse foreign influence on the Clydesdale horse. We believe, however, that this cleanliness and lightness have rather been due to carelessness and ignorance on the part of a portion of the farming community themselves, and also to a craze that at one time set in for clean legged horses, and the result of which time only can eradicate. During the progress Westward of the original stock, it, doubtless, was sullied at different points by the infusion of light blood by the way, and the roans that are so commonly found in Oxfordshire and portions of Cheshire would seem to indicate that, because, although in them very often the weight of body exists, we find a tendency to light and cleanly legs, so much so that when placed alongside animals bred in the Fens, it would be hard to convince one that they, at any time, could have been connected.

The success that has attended Shire horse breeding within the counties of Derbyshire, Staffordshire and Leicestershire, is not in the least degree hard to account for. As before pointed out, we turn to Volume i. of the Stud Book, and at once find evidence that pedigree and breeding have there been valued at their true worth. Breeders evidently implanted the original Lincolnshire stock on to the Derbyshire limestone, carefully preserved their mares from outside contamination, and what do we find as the result? Why, a source of genuine

Shire purity, a fountain head, as it were, to which the breeders of this kingdom gladly turn to replenish their studs. Unfortunately for the present generation, Derbyshire records are somewhat involved in the mysteries of the past. Traditions once handed down with some semblance of care have, in a great measure, been lost and forgotten, till now apparently only a few fragments remain. Only within the last year or so there went over to the great majority one of the oldest links between the present and the past, in the shape of an old stud groom, whose grandfather in his day was at the head of a famous stud owned by people of the name of Gallemore, who for generations had a celebrated Shire stud within two miles of Calwich Abbey. At the time when Prince Charlie marched on Derby in the famous '45 (1745), this old retainer was forced to take refuge from the invaders, and place the stallions of this stud in a place of safety. This he successfully did, and if curious readers will take the trouble to investigate Volume i. of the Stud Book, they will find several of the original Derbyshire stallions named Gallemore, having been doubtless in the first place christened after their owners. The stud was stabled at Croxden Abbey, and from its courtyard the horses went forth into hiding. Though it cannot be stated as an absolute fact, all the evidence points to the famous Packington Blind Horse having been begotten at this same place.

An ancestor of Royal Albert 1885 was another veteran that had a varied experience, for John Bull 1169 was rescued from the hands of the gypsies wandering by the road-side, and was in a very sorry plight. The father of Mr. Wright, late of Tideswell, Derbyshire, purchased this old horse for a mere song, owing to his having a thick leg, and Mr. Wright being a veterinary surgeon as well as a stallion owner, patched him up and travelled him. To this little incident how much do we owe to-day? Royal Albert doubtless would never have come into existence, and the excellence of mares of that noted strain would never have been known, but for that rescue.

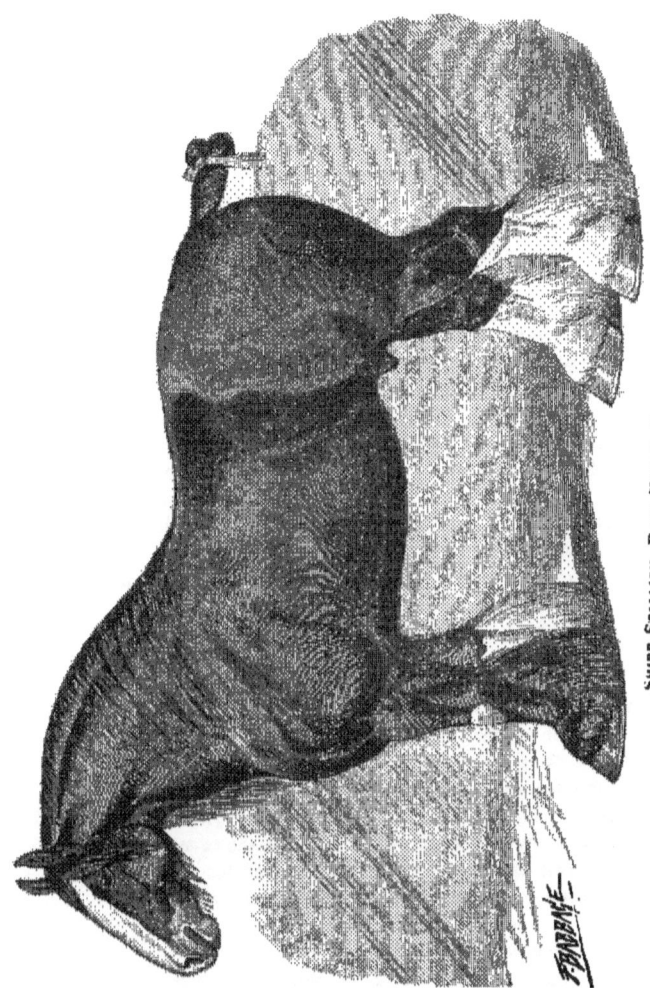

Shire Stallion, Bury Victor Chief, 11105.
Winner of Champion Prizes at London Shire Horse Shows, 1891 and 1892, &c.
The Property of Mr. Joseph Wainwright.

Lord Byron, the sire of Julian 3766 (who is with us still and is located in East Anglia), was the last stallion that the celebrated Chadwick family owned. They were quite among the best known stallion owners of olden times, inheriting the business from father to son as had been the case in almost every county in England. At the death of Mr. Chadwick, the horse Lord Byron, strange to say, was purchased by a tailor, who in turn sold him to a cloth hawker. Tailors, as a rule, are certainly not given to horse-dealing, but the fraternity in this instance, must have seen a source of profit somewhere in Lord Byron, otherwise he would not have long remained in the hands of the benchers.

These scraps of information regarding the past of our famous breed only make us wish for more, and when we consider what Derbyshire has done in contributing to the building up of Shires in other counties, and remember that for years it furnished both mares and stallions to all parts of Scotland, we can appreciate to the full the foresight of its breeders who have turned neither to the right nor to the left, but have stuck to the old blood through good and evil report.

Our Own Times.

We now propose to touch lightly on what has happened subsequent to the starting of the Stud Book, giving also a description of what the best types of Shire horse should be so as to come up to the requirements of to-day. As a very natural sequence to the registration of Shire horses, a great impetus was given to cart-horse breeding generally, and numerous studs were formed throughout the country, while prices rose rapidly. To the tenant farmer this has been a great assistance in one way, because during a period of very great agricultural depression, the Shires produced on the farm have been turned into money at exceedingly remunerative prices, and this, of course, has naturally assisted in warding off depression. Experience, however, seems already to have proved that eventually the value of all animals must depend on the marketable purpose for which they are bred. Thus, for instance, let us take the case of Shorthorn cattle. Have we not seen members of certain families selling for thousands of pounds; but as beef, butter, and milk may be called the marketable purpose for which they are bred, so their values have decreased until they approach the prices of their less fashionably-bred neighbours. It must be expected, then, that the Shire will go through a similar ordeal, and already it has been found that the market (at least for indifferent stallions) has been flooded, and disappointing prices have been the result; while at the same time the actual work horse has become wonderfully scarce owing to circumstances that now appear simple but certainly were not foreseen.

The large studs that have been established by noblemen and gentlemen have undoubtedly done a very great work and brought the breeding of the Shire horse down almost to a science, but in reference to these extensive establishments it may be questioned if their influence has been entirely advantageous, because, indirectly, the user of the Shire in the labour for which he is bred has practically been ignored, owing to the prohibitive prices that breeding animals have fetched. From this cause in a great measure has sprung the scarcity of work animals that of late has existed, and consequently a change is even now taking place, and the highly-bred brood mare is again returning into the hands of the tenant farmer, destined from henceforth to contribute a share of her progeny to daily labour.

But although the disadvantage above noticed has been the natural result of attempting to convert every individual Shire horse into a breeding animal, the benefits that have accrued infinitely outweigh the drawbacks. When one looks back some ten or twelve years to the first shows held at the Agricultural Hall at Islington, and compares the animals that competed then with those that do so now, the change is

Shire Mare, Rokeby Fuchsia. The Property of Mr. John Parnell.
Winner of Champion Prizes, London Shire Horse Shows, 1893 and 1894.

marked indeed, and this is undoubtedly due, firstly, to the efforts made towards the attainment of the best results by the Shire Horse Society; and, secondly, to the spirited manner in which the breeding community at large have entered into the work, and backed them up in every possible way. The result certainly is most highly satisfactory in many ways. Formerly unsoundness, round fleshy-legs and upright joints were the rule, while action was very indifferent. Nowadays, probably the improvement is most marked in the mares of all ages, and quality has unmistakably made great strides to the front in all the female classes, while an endeavour has also been made to engraft an equal measure of quality on the stallions, and for this reason we find that masculine sires are inordinately scarce. If the Shire horse of the future declines in character it undoubtedly will be due to this cause, viz., a tendency on the part of breeders to use stallions not sufficiently masculine in type, and to lean too much to smartness, cleanliness, and that fatal quality that some applaud, "prettiness." It would seem evident that quite sufficient advance has been made on the side of quality; the Shire mare has enough slope in her pastern to soften all jar entailed in work; her freedom from side bones is very marked when compared with pre-Stud Book days; her hind leg has been studiously improved, and now care only has to be exercised to avoid converting some of the improvement into the sole requirements of the Shire mare. As to the stallions, we would also give a word of warning. Many people are only too apt to believe that what are excellencies in the mare are necessarily also merits in the stallion. No greater mistake could be made. It is essential that the mare should have plenty of depth in her ribs, and general roominess in her middle piece, as her chief function is to carry her offspring during its period of development, and ample room must be necessary for the proper maturing of her progeny. On the other hand, a stallion of this type, viz., with a tendency to excessive middle, is almost invariably a failure at the stud.

CHARACTERISTICS OF THE BREED.

The Shire stallion should stand 17 hands or over, his legs should be as big and massive as it is possible to obtain them consistent with flat bone, which should measure at least 11 to 11½ inches below the knee, and 1 inch to 1½ inches more below the hock; the hair should be plentiful at all seasons, not wiry, but strong and decided, without any tendency to woolliness. The action should be most particularly noticed in the walk, which should be straight, level and true, and should be the walk of a cart horse, forward and free, but partaking in no respect of the jauntiness of the nag. The hocks should at all times be kept together and in position. The feet should be wide and open at the heel, with wall of sufficient depth to avoid any resemblance to flatness of foot; the pastern all round should have sufficient slope to enable the machinery to work smoothly, but long and consequently weak pasterns are to be avoided. The head in the stallion is of vast importance. It should be thoroughly masculine in character, and all trace of "ponyness" should be studiously avoided.

With regard to the attributes of the best type of Shire brood mare, it is perhaps correct to say that breeders and judges are possibly somewhat at variance, because the mare that usually produces the best results to the breeder is but seldom the animal to catch the judge's eye in the show ring. On the female side great size, or in other words height, is not only not an essential, but is usually detrimental. The typical brood mare should rather be long, low and wide, standing on short legs, with well sprung pasterns and strong open feet, while the bone should be as wide and massive as can possibly be had, and the hair should be very abundant and worn at all seasons; the depth of both the heart and short ribs should be conspicuously present, and the walk should be true and level, without any symptoms of rolling; this latter point, of course, applies equally to both mare and stallion. This type of animal

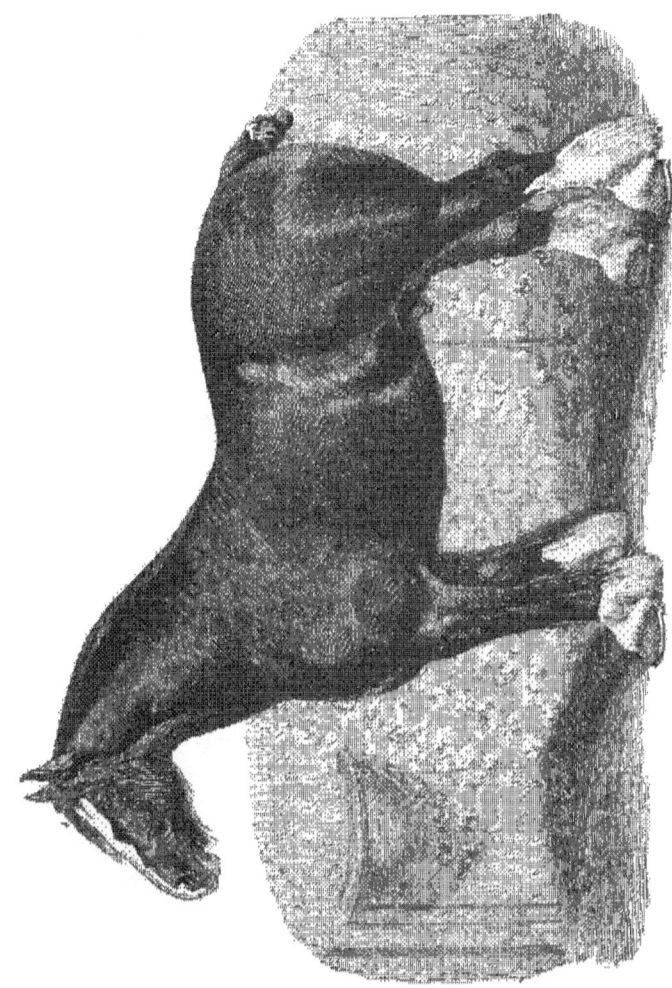

SHIRE STALLION, VULCAN 4145.
Winner of Elsenham Challenge Cup at the London Shire Horse Shows, 1889 and 1891.
The Property of the Earl of Ellesmere.

being somewhat unpretentious in character, and wanting possibly in gay carriage and dash, often fails to find favour with judges, but the breeder of experience will have little difficulty in selecting such animals for the purposes of their studs.

Of all the various influences that of late years have affected the Shire horse, probably no one thing has done more to stimulate attention to the breed, and to encourage its development in every way than the special Shire Horse Show at Islington, and also the shows of country societies that have given liberal prizes to the Shire breed. The show *par excellence* of the year, and the one that breeders look forward to and attend with the greatest amount of interest and enthusiasm, is undoubtedly the gathering held annually in the Agricultural Hall, Islington, in the month of February, which from comparatively small beginnings has in late years almost exceeded the accommodation of the Hall; in fact, it has been found somewhat unwieldy, so much so that steps have been taken with the view of limiting the entries, and thus encouraging a higher order of average merit. Apart from this particular show, almost every corner of the land has in one form or another had its exhibition of Shire horses. One effect that these shows have had, is that a certain remodelling of type has taken place, and the judges doubtless have been accountable for such alteration.

In former days, as we have seen, our forefathers for years struggled to foster and encourage the weighty element. Within the last fifteen years, however, a gradual lessening of weight has to a certain extent taken place, owing undoubtedly to the tendency of judges at shows to encourage quality. That the horse of former days was in certain instances coarse and somewhat fleshy-legged is undeniable, but the result has been, as is often the case, that a proportion of breeders have occasionally gone to extremes, and bred for quality without regard to weight. The consequence of this innovation has been that the weighty element is becoming once more in demand, and signs are not wanting that a revulsion of feeling is again taking place in regard to this question, and that stallions that possess some of the characteristics of the old-fashioned type will be sought after, in order to cross with mares of quality. Another outcome of the prevalence of shows has been that a great desire has naturally been cultivated among breeders to gain distinction in the show ring, and this feeling has again stimulated the forming of studs in increasing numbers principally for showing purposes, which has entailed the maintaining of a large number of animals in a state of idleness. This, however, is one feature of the Shire horse question that has been of very doubtful benefit to the breed at large. The end and object of all Shire breeding must eventually resolve itself into endeavours to produce the type of animal that will be the most profitable to the farmer, and experience has taught us that without weight we are nowhere, for if we present to the dealer the nicest turned and handsomest animal in the world, with sloping pasterns and all the requisites that of late years have been so much sought after, what do we find? Why, that the price offered for such in the open market hardly repays breeding, and but little exceeds half the price readily obtainable for one with the requisite weight. There was also another circumstance that inflicted temporary injury on the breed, and that was the excessive demand for Shire stallions that for several years existed. This trade was "boomed" after the usual American fashion, and everybody jumped into it, expecting at once to become rich. While it lasted, certainly vast numbers of Shires left our shores, but it must be confessed that in many cases the quality of the animals so taken was quite of secondary consideration. A few importers then set up a certain standard of horse that was very far removed from the best class of Shire. This type of animal was cleanly legged, up-headed and flash. For such stallions fairly good prices were given, until some English breeders almost began to imagine that, after all, probably Englishmen were mistaken in their type and that the Americans were right. Those who took this view of the

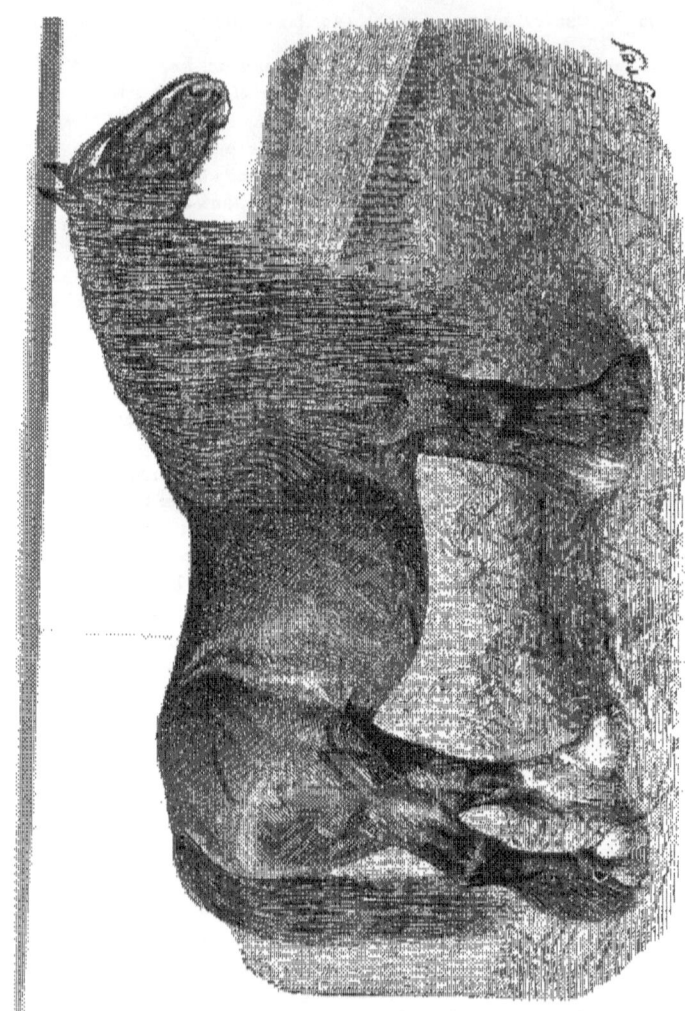

SHIRE MARE, LOCKINGTON BEAUTY.
Dam of Noted Prize Winners.
The Property of Mr. A. B. Freeman Mitford, C.B., M.P.

case, and acted on it, found themselves overloaded with horses that were almost unsaleable, and their condition was certainly not to be envied. A little reflection should have taught these gentlemen that one single stallion that is up to the proper English standard is worth at the very least as much as several animals of the other sort, while at the same time he is a benefactor to the breed at large instead of helping to deteriorate it. Englishmen are once more rapidly coming to their senses, and find that in breeding Shire horses they must not turn either to the right hand or to the left; they must not be carried away by passing fancies and fashions, but at all times and in every possible way endeavour to keep the Shire horse in the position that he occupies—at the head of all the draught breeds, by upholding weight, and by recollecting that what is estimable in a mare is almost invariably a defect in a stallion; that in selecting a sire true masculine character throughout must be at all times kept prominently in view; and that it is not always the horse whose showyard career has been most brilliant that is calculated to do the greatest amount of good at the stud.

Some Modern Sires.

Any sketch of the formation and growth of the Shire breed would surely be incomplete were the most famous of our present sires to be left entirely unnoticed. In undertaking such a description as we have indicated we find ourselves surrounded by many difficulties, for in the first place, it is, of course, an utter impossibility to chronicle every animal that perchance has begotten a winner and thus has gained notoriety; and on the other hand, the list might well become unreadable were we to include many of minor note, and should any such creep in, owners of stallions of similar merit might justly complain that they had an equal right to be brought prominently forward. For these reasons, then, it is believed that the safest and most equitable course will be to give a short sketch of the champion cup winners at the London Shire Horse Shows, making only one exception, viz., Lincolnshire Lad II. 2365, and the reason for doing so is that he stands alone as a sire, a grandsire and a great-grandsire of winners, while at the same time, although foaled in 1872, he still retains his health and vigour sufficiently to take his place in the stud and look around at his offspring down to the third generation, reaping the honours of the showyard thick and fast. Lincolnshire Lad II. 2365, was foaled just twenty years ago, and was bred by Mr. Fred Ford, Locko Park, Derbyshire, thus furnishing another illustration of the excellence of Derbyshire blood. This animal affords a curious instance of how one may be deceived by trusting to the eye alone in coming to a conclusion as to the merits of a horse as a sire, and his career also proves that it is quite possible to have as good a sire as need be and still not a Show horse. A grey in colour, the first point that strikes one about Lincolnshire Lad, is that although all along his top he is capital, yet his middle piece is decidedly light—a peculiarity that has very often been noticed as commendable in a stallion but fatal in a mare. Commencing once more at the ground, his next peculiarity is that he wears an enormous quantity of hair, and that it hangs and grows all down the front of his shins and fetlocks right over his feet, giving, at first sight, the impression that he is wanting in joint; but turn this hair back, and the mistake is at once apparent. Still, undoubtedly he is a peculiarly made horse, for he is not particularly full of muscle, and all over he shows decided narrowness; he stands, however, 17 hands high, and moves very well, and is full of courage and fire. We have thus closely described this horse because the points specially dwelt upon do not reappear in his offspring, but at the same time there is a general resemblance in his stock that make them by no means difficult to recognise, especially as regards the head, but the strangest thing of all is that the

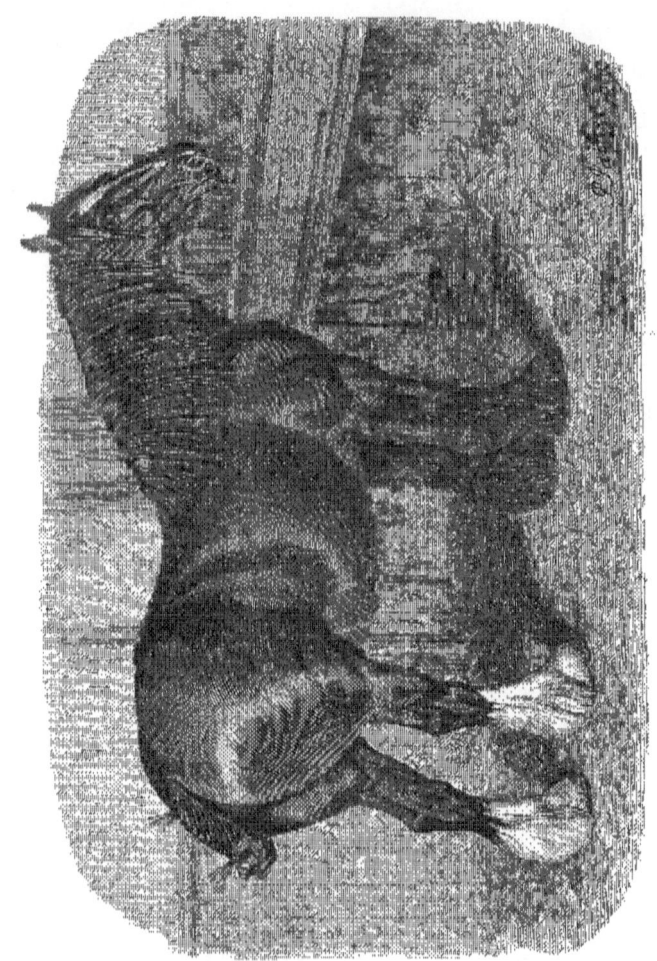

SHIRE MARE, STARLIGHT.
Winner of Lochinge Challenge Cup at London Shire Horse Show, First R.A.S.E., &c.
The Property of Mr. Fred. Crisp.

profuse hair on the shin only reappears in solitary instances, and a leading feature that follows the strain throughout is the excellence of the pasterns—a point for which a casual observer would say the old horse was not remarkable. His career has been a somewhat curious one, and in him is exemplified the difficulty that undoubtedly exists in ascertaining the true merits of a stallion when he is kept especially for the service of mares in any private stud, for Lincolnshire Lad in the prime of his life was quartered at Worsley, where he remained some time, but as a whole with rather disappointing results, although now his son, Lancashire Lad, promises to perpetuate his father's fame. The expectations that had been formed of him not having been realised while in the possession of Lord Ellesmere it was decided to part with him, and a purchaser was found in Mr. Walter Johnson of Hatfield, near Doncaster, in whose hands he has succeeded marvellously well, and moreover his stock, not developing early, began to come into prominent notice shortly after he passed into Mr. Johnson's hands. Among the most notable of his direct progeny are Harold in. the male line, and Scarsdale Bonny in the female, but although there are an enormous number of valuable horses and mares from him, his chief distinguishing merit would seem to be his power of transmitting good quality through several generations. These could be extended down in the form of a family tree, far more extensive than present space would permit us to give; and having already pointed out that Lincolnshire Lad is by no means faultless in conformation, it remains now to discover whence this extraordinary prepotency is derived. To find out this, however, we have not far to look, for by turning to page 259 of the first volume of the Stud Book we notice his sire K., *alias* Lincolnshire Lad, *alias* Honest Tom 1196. To this horse reference has already been made. He is familiarly known as Drew's Lincolnshire Lad, and his breeding has been traced. Still, we would point out that the accuracy of Mr. Drew's opinion of the breeding qualifications of Lincolnshire Lad 1196 is, in these days, being borne ample testimony to, by the doings of his son, his grandson, his great-grandson, and his great-great-grandchildren. Once more returning to the breeding of Lincolnshire Lad II. 2365, we find that his dam was by Matchless 1506 (Dan Howsin's), and so on down to Lion 1368, who was foaled in 1820. So that on both sides of the pedigree the blood is old and blue, and much of it has been in the possession of good men and good judges, both excellent recommendations as to character. Further than this description of the old grey horse we must not go, and possibly even in shortly introducing this account we may have transgressed the rule laid down, but as he seemed to hold a unique position, it would have been a mistake to have omitted reference to him. It may be added that Lincolnshire Lad II. was purchased in 1894 by Mr. F. Crisp, White House, London.

———

We shall now proceed to give a description of the winners of the champion cup at the London shows. Beginning with the year 1880, we find that this trophy was carried off by the Worsley stud, being placed to the credit of

Admiral 71

who was bred by Mr. Milner, of Kirkham, Lancashire, got by Honest Tom 1105, dam by British Ensign. Admiral was a dark bay horse with black points, as grand a set of hard flat legs as need be, and very flash hair; in appearance he was all over a very attractive horse, though in his younger days, perhaps, he looked rather light in his middle piece. He remained only one season at Worsley, and sired some useful, if not sensational, animals; shortly afterwards he was sold to Mr. Scott for exportation to Australia, at a very handsome price.

Spark 2497.

In 1881 the cup was won by Spark 2497, the property of Mr. W. R. Rowland, Creslow, Buckingham. Spark was an

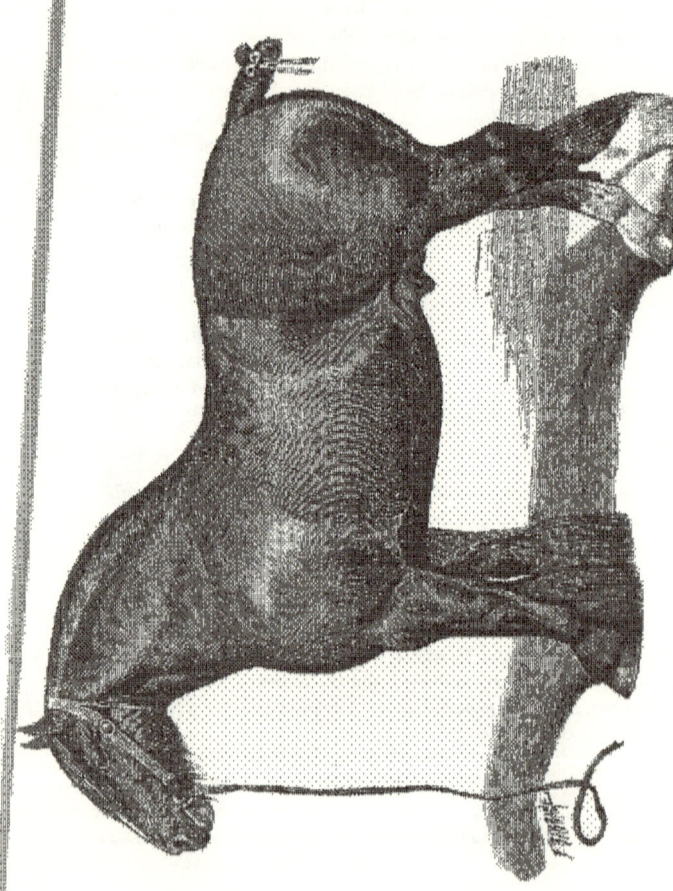

Shire Stallion, Dunsmore Willington Boy 13021.
Winner of Prizes, 1893 and 1894.
The Property of Mr. P. Albert Muntz, M.P.

entirely different type of horse to the winner of the previous year. He had commenced life well, as we find that as a foal he gained first at Wycombe, and that in the following year he was first at Aylesbury, and first at Buckingham. A massive, big, weighty horse was Spark—what might now be termed slightly old-fashioned in appearance, but one that showed himself capable of producing a very useful animal when well and suitably mated. His bone, possibly, was a trifle round, and his hair rather strong to suit fastidious tastes, but that might have been corrected in his produce. His champion win took place in his three-year-old form, and shortly afterwards he passed into Sir Walter Gilbey's hands, bringing what in those days was considered a sensational price, viz., 800 guineas. In 1883 he again succeeded in winning the championship for the Elsenham stud. One of his best sons was a horse belonging to Lord Egerton of Tatton, named Blue Beard 3472, a roan that had a great deal to recommend him; he ultimately went to America, where he distinguished himself in the show ring, and was sold for a very high price. Shellow Spark 3306, who was the property of Mr. John Rowell, was another of his sons that, as a youngster, promised well. In Spark's case it would seem that the second generation were destined to improve on the original, for a son of his named Royal Spark 4659, that went into Derbyshire, although in himself a rather plain horse, got some exceedingly useful stock, of which the best undoubtedly was Scarsdale Rocket 12249, a horse that very narrowly escaped being quite at the top of the tree; he was full of class, character and action, while at the same time he retained some old-fashioned locks from the knee and other shin characteristics that indicate a good, old, rough origin. Spark was by The Colonel 2701, and his dam was a most successful winner, as she carried off as many as eleven first prizes. She was by King Charles 1207, who ran back to Active 22. Spark was also the sire of some valuable mares.

Bar None 2388.

In 1882 the winner was Bar None 2388, and probably no horse of modern times has so successfully upheld his high reputation for a long period. Mr. Forshaw's horse we thus find sandwiched in between Spark in 1881 and 1883. This seems a somewhat strange coincidence, because Bar None was a totally different type of animal; he was all quality, but at the same time he was a big horse; his bone was beautifully flat and hard, his feather straight, silky, and plentiful, his joints, and the angle and side view of his hind leg being capital. He was foaled in 1877. When seen in his fifteenth year few people would have imagined that he was that age, so fresh and clean was he—in fact, only two years previously Mr. Forshaw seriously contemplated showing him once more in London, and had he come off with flying colours it would have been a marvellous performance in a thirteen-year-old horse. Bar None's fame, however, comes from his success at the stud; mares by him are sought for far and near, and they have undoubtedly one qualification that in a broad sense is invaluable, viz., that of crossing successfully with so many different strains. He invariably transmitted quality and "classy" legs, and his stock are easily picked out anywhere. He was sire of that very grand filly, Bar Maid, the property of Sir Walter Gilbey, who bought her at the Scawby sale for the highest price that had been paid for a female of the same age. Argosy and Challenge were two others that won at Islington and at the "Royal" Show, and besides there were the Worsley mare Golden Drop and Mr. Wainwright's Primrose. These are some of the most distinguished of his produce, but were we to attempt to complete the list we should occupy too much space. In stallions, probably as a sire Mr. A. Ransome's Hitchin Duke 9586, was one of the best; at the same time there are others well worth notice, such as Everton X L 5839, Headmaster 4448, Grey Friar 13127, and several more. Bar None was a Yorkshire-bred horse, having been

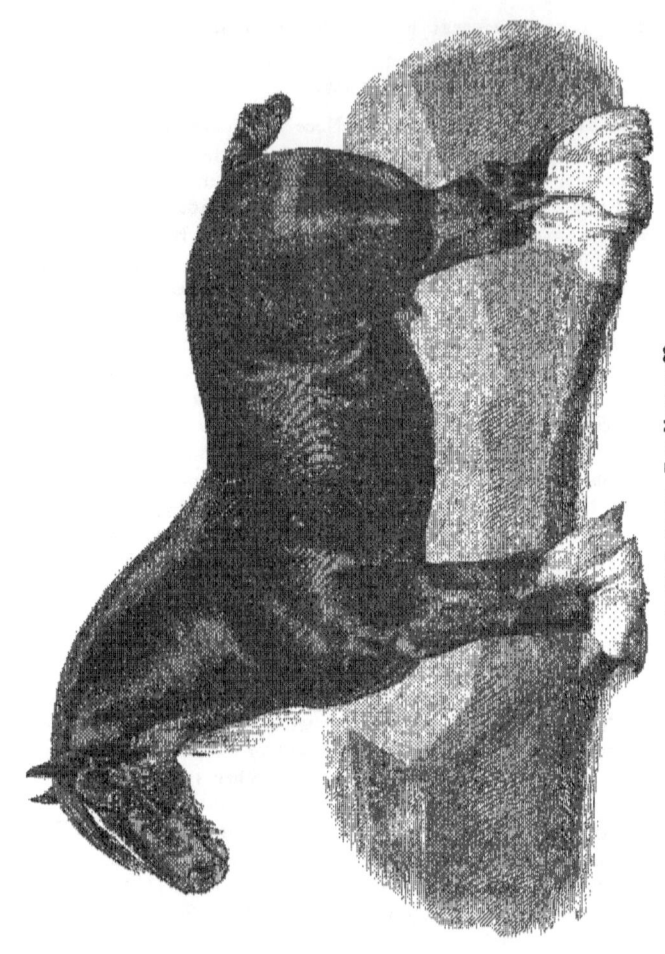

Shire Stallion, Bar None 2388.
Winner of First and Champion Prizes, London Shire Horse Show, 1882.
Owned by Mr. James Forshaw.

foaled at the farm of Mr. Holmes, Fenwick Hall, near Doncaster; he was got by Lincoln 1341, a wonderfully good weighty, brown horse, bred near Grantham, and a son of Enterprise 764. On the side of his dam he was still more Yorkshire-bred, as she was by Great Britain 973 (bred near Doncaster), and his sire was bred near Snaith. Undoubtedly Bar None's best qualities came from his sire Lincoln, and strange to say, although his dam, her sire, and grandsire were all roans, a roan by Bar None is never seen, the all-pervading colour being bay.

Enterprise of Cannock 2772.

The year 1884 brings us to Enterprise of Cannock 2772, a brown horse with white markings, and at the time he was exhibited the property of the Cannock Agricultural Company. The fact of this horse gaining the championship was another striking instance of how the tastes of judges differ, for we find the massive Spark winning in 1881, the high quality horse Bar None in 1882, Spark once more in 1883, and then still another reversion back again to quality, in placing Enterprise of Cannock at the top in 1884. As to the merits of this horse, opinions at the time widely differed. Undoubtedly, he was all over what is termed a "tasty" horse; he was wonderfully smoothly turned in all his outlines, his head and tail were set on high, and the former was neat, but not masculine; his bone was hard and clean, and he was certainly too devoid of hair; the weakest point of all being his knees, which were small, and, in fact, his foreleg altogether was wanting in substance. On the other hand he was distinctly a Show horse, for he went in marvellous form, and with wonderful force and courage——doubtless, great aids in the show-ring, and probably to these qualities he owed many of his successes. As a sire he was not very successful, but at the same time it must be recollected that his opportunities were few, because immediately after he had won at Islington he was sold, together with Harold and a black horse from the Mirfield stud, to the late Lord

Hindlip, to serve the tenants' mares on his Worcestershire estate. Owing to the death of Lord Hindlip almost immediately afterwards, he was put up to auction at "the Hall," and eventually found his way back to his old Cannock home. At this time the American trade was at its height, and during the succeeding winter he was purchased by Mr. Galbraith, of Janesville, Wis., but was destined never to reach his western home, as, unfortunately, he was killed at sea, together with many more, owing to the vessel encountering one of the most violent storms ever known on the Atlantic.

Prince William 3956.

Prince William 3956 was champion in 1885. This horse is the head of a family that is now famous. He was bred by Mr. Potter, of Lockington Grounds, near Derby, and sired by William the Conqueror 2343, out of the famous Lockington Beauty. Prince William won the championship at the early age of two years, and as in the previous year he was first in his class, and in 1886 again secured the championship, he may be said to have done what none had before accomplished. In type he varied considerably from Enterprise of Cannock; neither did he much resemble any other members of the same family. Undoubtedly, there is much about Prince William that indicates a sire; he is a strong boned and well haired horse, though his hind leg and joints could be improved. In motion, however, he at once catches the judge's eye, as his walk is true and good, and his action in his trot remarkably bold and free. At the time when he first gained the championship he was the property of Mr. John Rowell, Bury, Hunts., who had purchased him from Mr. Potter, after he won in his class the previous year. Mr. Rowell's judgment in selecting him proved to be quite correct, not only on account of his winnings, but also because he was able to sell him to Lord Wantage for 1500 guineas. His progeny at the Lockinge sale in 1894 were admired, and realized high prices. Among the most noted of his stock are Mr. Locke King's

Straiton Horse

British Flag III. 12841, and Mr. Bouch's black filly, that was first at the "Royal" in 1892, and won many other prizes.

Staunton Hero 2913.

In 1886 the Elsenham stud was once more at the top with Staunton Hero 2913, another son of William the Conqueror 2343, and bred in Derbyshire by Mr. Chappell, who sold him as a two-year-old to Mr. Douglas for exportation to Canada, but as already indicated he passed into the Elsenham stud. Staunton Hero was a brown horse, not standing quite high enough to compete in the big class, a very truly set one, with high-class legs and nice, straight, plentiful hair—in fact, his character was very much such as one would expect the stock of "Ould William" to be. His list of show-yard successes was large, and he won in his class once again in London subsequent to his gaining the championship. After standing some years at Elsenham he was sold at auction at one of the sales there, when the Duke of Westminster secured him for the Chester district, where he is understood to have proved a valuable sire, and to have crossed very well with the mares of that district. The celebrated mare, Dunsmore Gloaming, champion at the "Royal" in 1893, is from a Staunton Hero dam.

Harold 3703.

1887 introduces us to the now famous Harold 3703, bred by Mr. Potter, of Spondon, Derby, and sired by Lincolnshire Lad II., dam by Champion 419. Harold has been already mentioned in this sketch of London winners as passing into Lord Hindlip's stud, together with Enterprise of Cannock. That was in the year 1884, when he was three years old, at which time he was the property of Mr. C. J. Douglas, who intended to export him to America, but instead of doing so sold him to Lord Hindlip. Harold's position in London as a three-year-old was not a high one as he got only highly commended, together with a number of others. Like several horses that have been referred to, Harold had his defects, particularly in his three-year-old form, but afterwards the principal fault has always been more apparent than real. To the casual observer he seemed rather short of joint, but when examined, this, it was seen, was caused by the old Lincolnshire Lad characteristic of wearing his hair down the front of his shin and ankle. These are Harold's strong points: first and foremost he is a thorough stallion all through, and there is no possibility of mistaking him for one of the weaker sex at any point; secondly, his commanding size and tremendous bone; and thirdly, in action he is especially good when leaving you, and impresses one with the idea of never-failing courage and fire. These qualities he certainly transmits to all his produce in a most remarkable degree, and any one to-day possessed of either a Harold mare or stallion can find customers by the score, while he possesses the rare qualification of getting mares and stallions equally good. At all the leading shows it would seem as if this strain had almost a monopoly of the prizes, as at the conclusion of any important gathering it is nearly always found that Harold blood has carried off the greater portion of the spoils. He now belongs to Mr. A. C. Duncombe, Calwich Abbey, Ashbourne, Derby.

Vulcan 4145.

1889 brought forward a horse to which rumour had beforehand assigned a high position. This was Lord Ellesmere's Vulcan 4145, bred by Mr. John Whitehead, Medler Hall, and got by Cardinal 2407 out of Jessie by Sir Colin 2022. Vulcan had been a "dark horse" up to his appearance at the London show, having been purchased by Captain Heaton from Mr. Shaw of Winmarleigh, who purposely had kept him in the background. Vulcan is a different type from the last-mentioned horse; in make up and appearance he is quite a Show horse, and when in the ring attracts great attention owing to his handsome, level appearance, and his flat bone, his

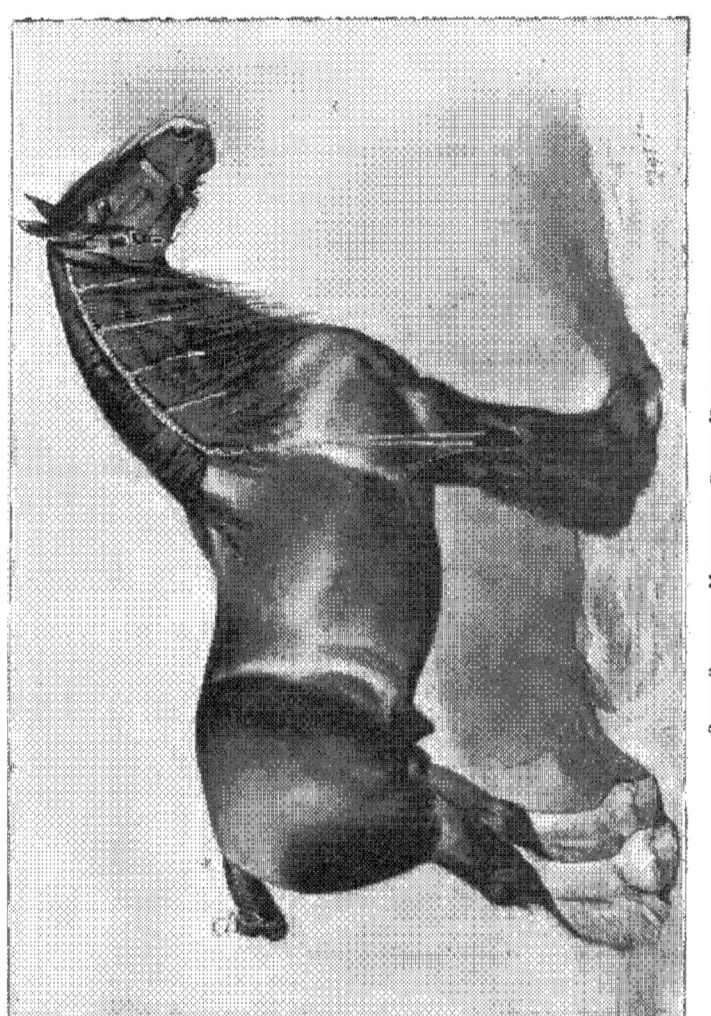

SHIRE STALLION, MARKEATON ROYAL HAROLD 15225,
Winner of First and Champion Prize at R.A.S.E. Show at Leicester, 1896, &c.
The Property of Mr. Alexander Henderson.

grand feet and pasterns and profuse feather. Differences of opinion exist as to his action, as he appears not quite to have sufficient firmness in motion, neither does he quite keep his hocks together. These may be called his weak points, while on the other hand he has a most attractive appearance and taking style that at once arrest attention. One special feature about Vulcan is his evident soundness, and the impression he gives one of his being a lasting, wearing sort. Probably the best female that this horse has sired is the roan mare, Dunsmore Fashion, that brought in many prizes to Mr. Muntz's stud. In stallions it is rumoured that a son of Vulcan out of Princess Louise is a marvel, combining size, bone, hair and action in a greater degree than any animal yet bred at Worsley.

Vulcan had the distinguished honour of carrying off the Elsenham cup in the year 1891, as well as in 1889, thus making a win outright.

Hitchin Conqueror 4458.

This horse secured the coveted position at Islington in 1890. He was bred by Mr. George S. Shepperton, Lockington, Derby, and was sired by William the Conqueror 2343, out of Flower, by Honest Prince 1058. He perhaps as closely conforms to one's idea of a big massive Shire stallion as anything in the list, and succeeded in being prominent in London more than once. From Mr. Arthur Ransome's hands he passed to Mr. Freeman Mitford, M.P., and has ever since remained at the head of affairs at Batsford. He stands 17.1, on short legs, though in his younger days he showed a bit of daylight; his bone, hair, and the way he is planted on the ground are all one could wish, but in the show-ring he does not appear to make the most of himself. Hitchin Conqueror has got some very useful stock at Batsford, though the district is against a horse distinguishing himself. In stallions of his get, the brown three-year-old I'm the Sort II. 7437, that was purchased by the late Mr. Punchard from the Cannock Agricultural Co., was undoubtedly up to that time the best. Then we have that massive son of his, Mars Victor 9889, out of Lockington Beauty, and now Mr. Mitford has a most promising one by him out of Madrigal, by Premier.

Bury Victor Chief 11105.

1892 brings us beyond the Elsenham cup once more to a challenge cup, which, however, was offered subject to the same conditions. Bury Victor Chief 11105 was the Reserve No. in 1891, when Vulcan carried off the honour, and was naturally looked upon as dangerous before the contest came off. This horse was bred by Mr. John Rowell, of Bury, Hunts., and sired by his own stallion, Prince Victor 5287, dam by Chatteris Le Bon 3023. Bury Victor Chief is a black, somewhat gaudily marked with white. As a thorough specimen of the Shire horse, correctly balanced and truly made, nothing could have surpassed this young stallion in his two-year-old form. After being reserve for the championship in London in 1891, Bury Victor Chief was kept back for the Royal Show at Doncaster in the same year, when he again won easily, and here the highest price hitherto given for a Shire was paid, when he passed into Mr. Wainwright's hands at 2,500 guineas. He became the London champion in 1892. Bury Victor Chief's strong points are his correctness of conformation, his levelness, as opposed to patchiness, all through his body, the marvellous evenness with which the muscle in arms and thighs is distributed, as it drops gradually into the big, hard, flat legs that show no weak point anywhere. The grandly developed two-year-old stallion seemed in the spring of 1894 scarcely to have grown on as one could have hoped, as he had then hardly sufficient length, nor was his action all that one could desire. But he again won the champion cup at Islington that year, and no other horse has distinguished himself so much as he has done at the London shows, having

been first there in his class in four consecutive years, and having been twice champion.

Rokeby Harold.

In 1893 the champion prize of the Islington Show was awarded for the first time in its history to a yearling, Rokeby Harold, and it was conceded on all hands that never in the history of the Shire breed had such a phenomenal yearling appeared before the judges. He was bred by Mr. A. C. Rogers, Prebend House, Buckingham, and was sired by Harold 3703, dam Poppy by Morning Star 1539. Rokeby Harold possesses all the qualifications of the typical Shire stallion, as he has size, colour, substance, bone, hair and perfect action, besides showing throughout perfect masculine character. In 1894 he very easily defeated all comers up to the final award for the championship of the show, which he contested with Bury Victor Chief. The judges for one half-an-hour were divided, and only at last gave way to the greater maturity of the older horse. This is the only occasion on which Rokeby Harold has suffered defeat, and he has all the indications of continued development up to his fifth year. His owner, Lord Belper, is certainly to be congratulated on being the possessor of so promising a stallion, and one likely to perpetuate the characteristics of the true Shire horse.

In writing this short sketch of the origin and progress of our greatest English draught horse, we have endeavoured to show that he is no recent upstart, but has existed in greater or less degree for at least some 2,000 years. He has had his ups and downs of fashion; breeders have at different times in certain localities, endeavoured to create what they supposed would prove to be improvements, by introducing alien blood to a breed that was well founded and established, but all such attempts have invariably proved anything but a benefit, and therefore it would seem necessary that any animals that show the slightest tendency to "cleanliness" and smartness should be carefully excluded from the ranks of breeding stock. If this rule is generally practised, we shall no longer hear any complaints of scarcity of weight or deterioration of the Shire horse, whose future is undoubtedly well assured, provided that all interested in the breed keep one, and only one, end in view, viz., the original purity of the race.

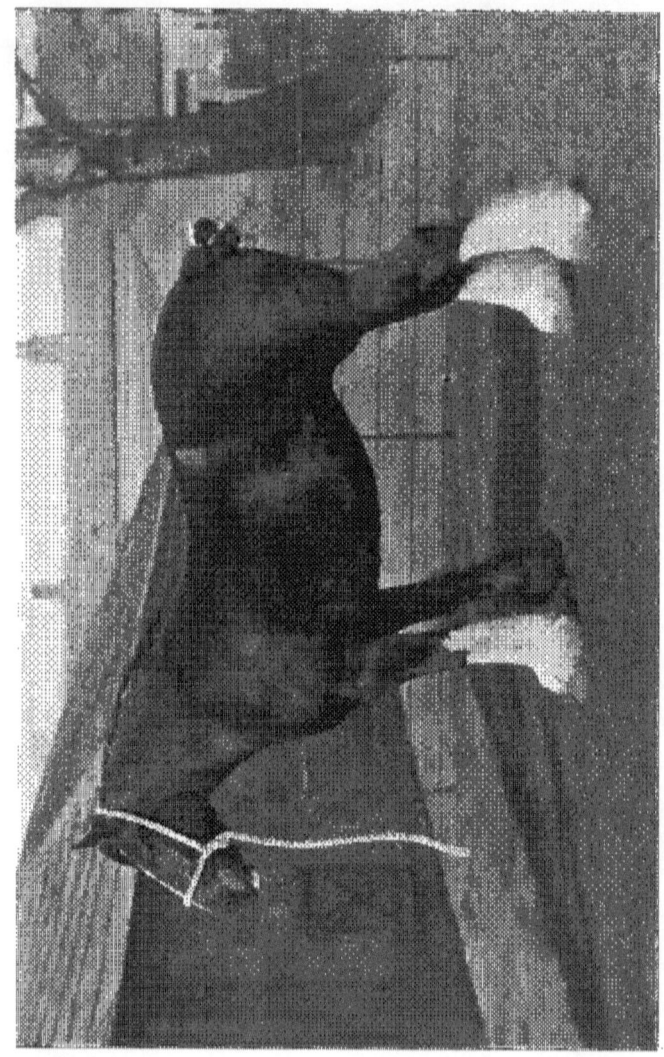

[From a Photograph by F. Babbage.]

SHIRE MARE, QUEEN OF THE SHIRES 20686.
Winner of Champion Prizes as best Mare at London and "Royal" Shows, 1897.
The Property of Mr. A. Grandage.

CHAPTER II.

THE SUFFOLK HORSE.

BEYOND the brief incidental remarks found in the topographical works on the county, no attempt to complete a history of the Suffolk horse appears to have been made previous to the year 1880. Young, Cullum, and one or two others mention the breed, and although what is found in their writings is by no means unimportant evidence of antiquity of origin, they seem to have recorded little more concerning the subject than their own personal observations supplied.

That ample material for such a history existed, the Suffolk Stud Book Committee proved beyond doubt, and in their first volume is recorded in consecutive form an account of the breed for some 160 or 170 years, with verified quotations carrying the history as far back as the early part of the eighteenth century. Limited in the district of its origin, and strictly local in its early development, there was no difficulty in getting at the historical facts which were in existence, if only time and means were forthcoming for the work.

The Stud Book Committee were singularly fortunate. The proprietor of the *Ipswich Journal* has an uninterrupted file of that county paper from the year 1720, and this was placed at the disposal of the then editor of the Stud Book (who also writes this account) for search and extract. Rich in news and notice of all matters connected with agriculture, frequent allusion is made to the Suffolk horse, both in this and another county organ—the *Suffolk Chronicle*. The native breed of cart horses appears even at that early date to have been a marked feature in the agriculture of the district. In the compilation of the history, recourse was had to advertisement cards, sale announcements, records of Michaelmas auctions, and, later on, the catalogues of the Suffolk Agricultural Association and the Royal Agricultural Society of England, and what appears to have been a more prolific source of reliable information, the verbal accounts and descriptions from the oldest grooms and horsemen in the county. The memories of some of these reached into the later decades of the last century. Twenty years' tabulation and preservation on the part of the editor of every scrap of information thus acquired, the voluntary help of all the breeders, with the ample funds at the disposal of the Stud Book Committee, enabled what has since been enchartered as the Suffolk Horse Society to place in the hands of the public a tolerably complete history of the native horse of the county of Suffolk.

So far as the origin of this breed of horses is concerned, little can be said. He appears to have been as indigenous to the eastern part of Suffolk as are the blue-black beasts to the Welsh hills, or the wide-horned, woolly-coated Highlander to the mountains of Scotland. Care and selection have modified his character, as they have modified that of other domesticated animals; but as regards his marked characteristics, few breeds have so tenaciously reproduced their salient features of identification as the original race of Suffolk horses. It is perfectly clear, and there is reliable evidence of the fact, that many of the most decided points which distinguished them two hundred years ago are rarely absent in the Suffolks of the present day. The short legs, the roomy carcase, the sorrel colour, the constitution, the length of days, and that inexhaustible perseverance at the collar are still prevalent features in the chesnut of our own time.

As far back as the middle of the last century, allusion is made to the purity of the breed; advertisements of that time going back for three generations, and noting that a horse of that date was "the truest bred cart horse in Suffolk," are frequently met with. In fact, in the earliest numbers of the county papers already alluded to, horses are spoken of as of the true breed of the county, with no more explanation of its origin or intimation of recent introduction than would be found in advertisements of the present day. Writers of our time have mentioned the fact that there is an element of Flemish blood in the early forefathers of the present race of Suffolk horses. Beyond the fact that a former owner of Holkham had a couple of Flemish horses, no record of any such introduction seems to be known. The only authority for even this fact is that there are portraits of two such animals in the family mansion. If these were used on the estate, the impression could only have been of local effect, and we may safely conclude that its influence was of no more permanent nature than other introductions of extraneous blood which, through accident or experiment, have had a local trial. The most recent known introduction of this element is in the pedigree of the granddam of a celebrated Suffolk horse foaled in 1846, and this mare must have been foaled some sixty years ago. Nothing nearer than the fourth or fifth generation from this horse can be alive at the present time. Supposing no other alloy is in the pedigrees of his living descendants, not one part in a hundred of their constitution can be Flemish. From the researches of the editor of the Suffolk Stud Book, the amount of Flemish blood in the present generation of Suffolks may be safely set down as practically *nil*. Of the introduction of other blood exhaustive notices are there given, but before these alloys are dealt with it may be worth while to give a direct quotation in allusion to the old breed.

Writing on this subject, and alluding to a period long prior to Young's time, the editor of the Suffolk Stud Book says :—

"There were draught horses peculiar to the county, and of standing enough as a distinct breed to maintain their prevailing characteristics through generations of descendants long after the original type had been considerably modified by repeated selection, and the introduction of incidental crosses. How long prior to Young's time the breed had existed we have no evidence to show. Improvement had been effected even during his life, so that when Sir Thomas Cullum's work was written, they were occasionally used for 'carriage purposes.' A few more years, and Jery Cullum records them as good movers. Ten years later, Sir Robert Harland's sale took place, where as much as £140 for a brood mare, and 40 guineas for a foal, were recorded. From that time detailed descriptions of individual horses of note, taken down from the lips of dependable witnesses whose memories extended into the last century, are now extant."

The improvement here spoken of was on the animal of Young's earlier inquiries. Further on, from the same authority, we learn that "Young's report of the agriculture of the county dates from the end of the last century (nearly a hundred years ago), and if his knowledge of horses was in keeping with his knowledge of other branches of the subject, the following description enables the reader to form a tolerable notion of what the animal was which has since developed into one of the most popular breeds of the present day. 'Sorrel colour, very low in the fore-end, a large misshapen head, with slouching heavy ears, a great carcase and short legs, an uglier horse, as the author says, could hardly be viewed.' Quite so; and the chesnut of forty years ago (written in 1880) still retained traces of these unsightly points, enough in many cases to identify his form with that of his ancestors forty years before that."

Now Arthur Young was a middle-aged man in 1780, and his memory would have carried him back into the first half of the century. He speaks of remembering the "old breed." The expression, the 'old breed," has a marked significance, and in following up the history of the Suffolk horse explains much that bears on the question of the purity of the origin of the race of Cupbearers, Foxhalls, Wedgewoods and the rest of the Suffolks of the day.

Americans, colonists and foreigners never fail to notice the uniformity of character of the Suffolk horse. There is something in the colour, style and outline, varied, but never obliterated, which speaks of a common origin. Whatever the objections to a Suffolk horse may be, no one denies the marked type of outward appearance he invariably exhibits. That there have been infusions of extraneous blood, the history, as given in the Stud Book, shows plainly enough. The fact is neither concealed, slurred over, nor doubted by the editor. But one thing is proved, and that is, that not one of the introductions from outside the county, not one of the strains of alloy in the male line, could stand before the influence which the old breed asserts. The produce of the crosses in some cases stood for years, but sooner or later they died out, and at the present time there is not a living Suffolk horse which is not descended from the old breed. But this fact has a still more extraordinary side to it. Incredible as it may appear, there is not a single specimen of the breed now in existence which is not descended from one single source of ancestry. Every Suffolk in the showyard, homestead, or the breeder's stable, is the lineal descendent of a certain horse of the "old breed"—a nameless sire, foaled in the year 1768, and advertised as the property of one Mr. Crisp of Ufford—a village three miles north of the town of Woodbridge. Nor is this a mere assertion, or the result of a fair conclusion from reliable data; it is a proved fact, which the pedigree chart in the Stud Book proves beyond dispute or doubt. Every link in the chain is clearly traced and identified, from the well-known winners at our summer shows in the year of grace one thousand eight hundred and ninety-three, back into the first decade of the reign of George II.—"a period," as the author of the Stud Book says, "not much on paper, but a glance at what has taken place during those years gives some idea of the time which the character of the present race of Suffolk horses has taken to develop. When the old horse was foaled George II. had been dead only eight years. The United States of America were then but a colony, whose weal or woe was at the mercy of some functionary at Whitehall. France had still twenty years of slumber before she awoke to the horrors of the Revolution. Snipes abounded within a few hundred yards of Trafalgar Square, and the route from Saxmundham to London, for common folks, was by a hooded waggon and six horses, and the passengers, so tradition says, wrote home from Melton the first night to tell their friends that, so far, the journey had prospered. The eastern part of Suffolk was little more than a sea of heath with sheep tracks, which are now good flint roads. Swedes and mangels were then unknown as agricultural produce. The Norfolk sheep was the only kind found all over the district, and no beast was made fat under four or five years old. But the Suffolk farmer had his Suffolk horse, and through all the years which saw the long dreary war with France begun and ended; Pitt, Fox, Trafalgar, Waterloo, wheat at £4 a coomb, in 1812 and the sad times of 1822, the first Reform Bill, the old Poor-Laws, the Repeal of the Corn Laws, the Crimean War—through all the times of which these events are but the landmarks of the history they comprise, down to the Show of the Royal at Kilburn, the breeders of Suffolk horses have been true to the native stock which their forefathers left them, but upon which they have gradually stamped their improvements, and while retaining the characteristic points of usefulness which 150 years ago had made them famous as a breed of English horses, have produced the animals now seen in the Suffolk classes at our annual summer shows."

Allusion has already been made to the description of the "old breed." Advertisements of a century and a-half ago recommended them to "breeders of good stock" for coach or road. The significance of this recommendation must be taken in connection with what the roads were at that time. What is now the old turnpike from Ipswich to London was then but a natural way, which traffic had selected as the path offering the least impediment to travel, but which had no more artificial foundation than the camel route across the Sahara. What was required for "coach or road" in those days implied a strong active animal, less cumbersome than the old black English draught horse, but which could, and willingly did, master an unusual depth of sand or other obstructive matter in the well-worn ruts, and where the roads were better, could maintain a six-miles-an-hour trot without undue strain on legs or lungs. The sixty-eight miles from Ipswich to London took three days to accomplish, and the six or eight sorrels which formed the team bore little likeness to what after the days of Telford carried the traveller over the same route. Of such was the old breed referred to by Young.

The introduction to the Stud Book (written in 1879) thus speaks of the old Ufford horse: "The first notice we get of a horse of the old breed, of whose undisturbed identity there is printed record, belonged to a Mr. Crisp . . . . he was the grandfather of the present generation of that name, and then for more than a hundred years the family seems to have been foremost among the breeders of Suffolk horses. The advertisement appears in 1773. The following year he is described as 'a fine bright chesnut, full 15½ hands high, with the additional notice that his owner 'has no occasion to say anything more in praise of him, as he is so noted a horse for getting fine colts;' and moreover, 'those who were unsuccessful with the said horse last year can have the use of him this season for five shillings.' His route appears to have been in the district of Woodbridge, with excursions to Saxmundham and Framlingham, 'so to continue the season, God willing.'"

Thus much of the "old breed." Of the various introductions of outside blood the Stud Book gives detailed particulars. The first of these, the most important, the last to die out, and

the most potent in effect, was a horse which belonged to a
Mr. Blake, who then resided in the parish of St. Margaret's,
Ipswich. The issue of this experiment were known as the
Blake strain. Blake's Farmer, as he was called, was a Lincolnshire trotting horse, a short-legged chesnut, with a pedigree
Lincolnshire throughout, with no Suffolk blood in it. He
was in the county fifteen years, and when the cross had been
tried, Blake advertised the fact that "many of his horse foals
out of cart mares were sold for twenty-five and thirty guineas."
It appears that Blake had four generations of these horses,
presumably each getting more Suffolk blood in his veins, and
less of the Lincolnshire trotting horse. The fourth generation produced Young Briton, foaled in 1796. He was the
horse "Squire" Wakefield bought for the Burnham district
in Essex, and is identified as the very animal alluded to by
Arthur Young as a Suffolk horse, for which his owner (the
said Squire Wakefield) had refused 400 guineas. So fashionable had the Blake horses become, that this Burnham horse
was one of twenty which at various times were advertised in
the *Ipswich Journal*. Indeed the old breeders, with whom the
writer of this article was intimate thirty years ago, all spoke
of the Blake horses as the most popular strains of Suffolks
when they were young men. This would be at the commencement of the present century. If not the most popular,
they were, at any rate, the most numerous. To all appearance they had gained a permanent footing as an established
branch of the Suffolk breed. Among the Blakes fifty years
after they were introduced we find more than one winner at
the Royal Agricultural Society of England shows. When the
meeting was in Wiltshire, Mr. Crosse's Shrewsbury Briton
won the first prize as the best horse for agricultural purposes
against all comers, Suffolks, Shires, and Clydesdales. Nineteen years after that, in 1864, the late Sir Edward Kerrison
won the first prize for Suffolks at the Newcastle meeting of
the Royal Agricultural Society of England with Ploughboy—
a sixty guinea purchase as a foal from a breeder in Essex,

who liked the sort from over the border even for the stiff clays of that stiff county. Both these horses were lineal descendants of Blake's Farmer. Twenty years after Ploughboy won at Newcastle the tribe was extinct; there was not a Blake in the county of Suffolk. They became extinct from no assignable cause; they did not go out of fashion, for Shrewsbury Briton was a charming horse, and Ploughboy had bone enough for a dray. It was the old tale over again, the fresh introduction vainly striving against the power of the old parent stock—a fact of which the annals of breeding again and again give the proof. It is simply this, that the power of assertion, the extent, the tenacity of retention of its characteristics, is in proportion to the antiquity and the purity of origin. From the year 1780 to the year 1880, just the 100 years, the Blake strain was in existence. Steadily making its way for thirty years, in the ascendant for thirty years after that; in thirty or forty years more the tribe was extinct.

Running side by side with the old breed, intermingling in after years, the lineal descent gradually disappeared; but the influence had a permanent effect. The description of the old breed is well known. The handsome fore-end, the activity, the neater outline, came in with the short-legged chesnut trotting horse, which Blake introduced.

The next important infusion of extraneous blood came through yet another Lincolnshire horse, or, at any rate, a horse which was bought in that county. This was Wright's Farmer's Glory, sometimes known as the Attleboro' Horse. He was a clean-legged chesnut, and as an old horseman, from whom the writer had the description, said, "he might have been a half-bred Suffolk." From the description given he probably was. At one time, between 1803 and 1807, there were eight of the Farmer's Glory stock at the stud. Four generations later we find Nunn's Boxer, foaled in 1821, credited with seven sons for service between 1824 and 1830. This was before the days of agricultural shows, and it is not till several further generations onward that we meet with names of this strain as winners in the show-ring. Chelmsford Champion, the property of Mr. G. Mumford Sexton, who had not then transferred his love to the Shire-breds, took the first prize for two-year-olds when the Royal Agricultural Society held their meeting in Essex in 1856. Coulson's Royal George, another of the same tribe, but of a different branch, beat Shires, Clydes, and his fellow Suffolks, when the Royal met at Norwich in 1849. Barthropp's Albert was second at York, and his Hercules was first among the Suffolks when the county show was held at Bury St. Edmunds. Garibaldi was first as a two-year-old at Newcastle in 1864; and Lewis Duke, yet another descendant in the tenth generation, took the first prize as a three-year-old at the Suffolk show at Ipswich in 1869, winning the champion cup as well, and was sold to go to Australia then and there. This list by no means completes the catalogue of winning animals descended from Wright's horse, but it serves to show that the tribe had got a footing, and numbered many showyard celebrities, albeit with the exception of the last-named the quality of the best was not over high. But like the Blakes they died out, and a dozen years after Lewis shipped his colt to Australia there was scarcely one of the breed in existence. They are totally extinct now. They were introduced some forty years after Blake's Farmer came into the county, and died out about the same time as did the descendants of Blake's horse.

These are the principal instances of the introduction of outside blood. The effect of the second cross was of no value. The animals named were more after the style of the first of the tribe—tall, angular horses, and heavy boned, but with less of the compact neatness of the Blake's, and it is well they gained no more footing. There were, however, other introductions, but as these were more alien to the character of the stock they were grafted on, they died out far sooner than either of those named. The best known were the Shaddingfield stock. They date from about the time Wright's Farmer's Glory came into the county—about the year 1800

The Shaddingfield stock were the produce of a trotting horse, the son of a thoroughbred owned by one of the ancestors of the present proprietors of the Benacre estate. They were very distinct in character, much whited in the leg, dark chesnuts, with thin shoulders, hard, clean limbs, and light of heart, with spirits that took them through many a day, which would have killed a hairy-legged heavy horse. They lasted but seven generations. They were in the land for nearly fifty years, the last of them being Mrs. Catchpole's Proctor, foaled in 1846. Old Moyse, an authority frequently quoted in the Stud Book, gives a quaint description of their origin. The first of the tribe was Barber's Proctor. Moyse tells the editor: "The sire of Barber's Proctor was Winter's Stormer. The dam of Barber's horse was a chestnut mare. Barber's was originally intended for a riding horse, but they broke his tail off when they nicked him, and he was then travelled as a cart horse. Winter's Stormer was a trotting horse of great substance; he was a son of Gooch's blood horse, brother to Thunderbolt." Mark the tale of these broken tails. The riding horse, to make him in fashion, was nicked, and this gave the upward turn we see in the hunting pictures of the last century. The Suffolk cart horse at that time was what was called "bung tailed"—the dock barbarously cut off close to the hind quarter. The accident in nicking Barber's horse made it necessary to amputate the tail altogether, and then, as Moyse tells us, "he was travelled as a cart horse"! Then follows a curious note in the Stud Book, singularly corroborative of the information furnished by old Moyse. "All this," says the editor, "occurred sixty years before Moyse related the story to me." That he was correct about the breeding of the sire of Barber's horse there is little doubt, for in the year 1789 one P. Winter advertises "a blood bay colt, full fifteen h., by Gooch's horse Stormer." He stood at Snape. Nothing is said in Winter's advertisement either of Gooch's Stormer being a blood horse or of his being brother to Thunderbolt; but the very year that he advertises his colt

Gooch advertises the blood horse Thunderbolt as brother to Stormer. On turning to Weatherby's Stud Book I find Stormer was bred by the Duke of Grafton in 1774. The age is exactly right, and what more probable than, if useless as a race horse, he should have been sold for a country stallion close by home? Now it matters very little whether Winter's Stormer was, or was not, the son of a certain horse, but it is of no little importance to test the credibility of an authority so often quoted as our old friend from whom this history comes. (Suffolk Stud Book, vol. i., p. 49.) Well might the Shaddingfield mares have their light hearts, wiry legs, and thin shoulders, for the pedigree of Gooch's Stormer, bred in the year 1774, takes us back over six generations of the direct male line, through Flying Childers to the Darley Arabian, brought from Aleppo about the end of the seventeenth century. Thence came the white legs, the dark colour, and the indomitable spirit that made the stock famous among the breeders in High Suffolk fourscore years back.

The stock had well nigh become extinct before the show-yard era had commenced. One of the tribe, however, was sold for 400 guineas to George IV.—" a sweet pretty horse" as old Moyse described him. Some of those Otley Bottom Proctors, as they were called (there were four generations of Proctors, all owned by Barber, of Otley Bottom), were heavy horses, dark, copper-coloured chestnuts, almost black, with gaudy facings, white legs up to the hock. The last of any note—one of which the writer of this article well remembers—was the property of Mr. William Wilson, of Baylham Hall, still to the front with colts of the highest class at the present day, and yet, strange to say, he was the owner of Proctor 67, whose sire was foaled in the year 1824. Nearly sixty years' uninterrupted ownership of Suffolk sires should give him a standing in authority that few can claim.

The mares and geldings of the Shaddingfield strain were smart walkers, thin in the shoulders, had upstanding fore-ends, with rare bottom, but they were light below the hock, with a

bit more daylight under them than the Suffolk should have.
But like the Blakes and the Farmer's Glory, they are clean
gone now, and their influence on the old breed was practically
nothing. The last actual lineal descendant was Mr. Manfred
Biddell's old Brag Mare 59, who died some years ago without living produce. The only reference to the strain which
appears in the pedigree of any horse of note, was in the blood
of Wilson's Goliath, a prize winner at the Royal Society's
Meeting at Norwich in 1849. His dam was by the last of
the Proctors, already referred to.

There is yet another infusion of extraneous blood, the circumstances of which are so strongly corroborative of the
theory already noticed, that it is worth mentioning. It is so
far remarkable, inasmuch as the horse travelled the county
nineteen years; had the run of the heart of the Suffolk headquarters; was under the charge of one of the best and most
popular horsemen of the day, and yet failed to make his mark
on the district where he was used. The fact is accounted for
by the violence of the cross attempted. Although a Suffolk
to look at, and travelled as such, he had apparently no Suffolk
blood in his veins. As affecting the history of the breed,
this introduction of strange blood is of no weight; but to
those who care to look into facts bearing on the theory of
breeding, the history of this horse is worth notice. He was
the son of a rough-legged, timber carter's horse, out of a
black blood mare. Such a cross was not likely to have any
effect on a breed of horses of at least two centuries' standing, but it is a singular proof of the theory that not even a
grandson is found, in advertisement, sale catalogue, or otherwise mentioned in any record, printed or verbal. The name
appears in the far-away branches of two mares on the Stud
Book registry, otherwise there is no mention of him in the
pedigree of any animal now living. Full particulars of this
horse will be found under Martin's Boxer 868, on page 71 of
the Suffolk Stud Book.

Before the institution of the Suffolk Stud Book Association, before any pedigrees were recorded, the Suffolk Agricultural Association offered prizes for the best horse for agricultural purposes, and although the competition very rarely indeed included anything beyond the prevailing chesnut, there was a feeling that it was time to have some restrictions as to breeding when the entry appeared under the county colours. The effect has been to render the breed secure against further experiment of engrafting other kinds on the native stock. As years go on the breed gets more consolidated, and even if a cross were tried now, in all reasonable probability the effect would be exhausted in two or three generations. On the Martin's Boxer cross the editor of the Stud Book remarks: "It seems as if the blood thus introduced was so foreign to the original stock that the parent stem rejected it, repelled it, refused to allow it to mingle with the ancient family strain upon which it was for twenty years persistently grafted. It reads as if Nature forbade the alloy. In the case of Blake and Wright, it seems as if the foreign element they introduced had some slight affinity, or at least some outward similarity, to the Suffolk stock. But in the case of Martin's Boxer 868, the infusion was so foreign to the nature of the old strain—the cross so violent—that it never gained a footing."

As regards domesticated agricultural animals, colour is generally a distinguishing element. The Welsh beast is invariably a black, with an occasional tendency to blue grey; but unless a cross is clearly traced, a red hair never appears. All shades and combinations of red and white, or red, or white, are met with in our pure-bred Shorthorn classes, but black is unknown. The West Highlander is of many shades of red, pale ochre or black, but white is never seen. No one ever bred a pure Hereford otherwise than red and white, and a Devon rarely varies his whole-coloured red coat.

As regards colour in horses, there is no shade known that has not been found in the English thoroughbred. There was a kind of blue roan that at one time distinguished many of what were termed Norfolk trotters; they have, however, been now so mixed with other Hackney strains that, like the thoroughbred, the high-stepping show Hackneys are as varied in colour as the ingredients of white, black, bay, chesnut and dun can produce. The Shire-bred man is in no wise particular. Watching the ring at the Royal, one sees black, brown, grey, bay and chesnut; with or without white, whole-coloured, blotched and sandy roan. The breeders of these fashionable horses on this point are totally without prejudice, and stopping short of sky blue or emerald green, they apparently claim all shades as the "true colour" of the Shire-bred proper. But the Suffolk man will have none of this. "Chesnuts all, and all chesnut, with white facings as few as possible," is his creed, and right bravely is this idea carried out. It is between forty and fifty years since a bay horse has been advertised as a Suffolk, and in this case it was acknowledged that on the dam's side there was a supposed infusion of other blood. When the Suffolk Stud Book Association was formed the committee rightly insisted on maintaining the original colour, thenceforth excluding every entry not of the recognised chesnut. But no shade of this one colour was objected to; nor does the white face or white leg act as an obstacle to registration. There are many shades of chesnut, ranging from the dark copper hue of Old Cupbearer, to the lightest sorrel. Crisp's Old Ufford Horse was bright in colour, but one of his sons at the stud was advertised as a dark chesnut. Forty years afterwards—in the fifth generation—Smith's Horse of Packam had a great run, and through him come all our Suffolks of the present day. He was a dark chesnut, "without white of any account." From that date to the day of Mr. Wilson's Old Times, the sire of the first prize yearling at Bury St. Edmunds in 1892, the dark chesnut has cropped out. In this case, in the paternal line, that colour lay dormant for eleven generations.

Those who remember so far back—they must be few now—must pass over Light Heart 1421, Hayward's Champion 680, Chillesford Duke 395, May Duke 426, Red Champion 435.

Capon's Duke 261, Catlin's Duke 296, Manchester Boxer 298, Catlin's White-Faced Boxer 299, Artess' Boxer 26, Groom's Ramper 636,—"a nice little pretty horse of common bright chesnut, a little white on forehead" (Moyse)—on to Brady's Briton 198 foaled in 1809, "a very dark chesnut, with a bit off his dock," before we again meet with that shade in the paternal pedigree of Old Times.

Old Cupbearer was, at certain seasons of the year, almost copper coloured; from him we get, through a third generation, a numerous tribe of winners in the showyard. Few of these are dark, and like the ancestry of Old Times, there are many generations before we meet with one of a dark shade. The effect of colour on sales is more than its effect in the prize ring; there, if of equal merit, the bright chesnut would have little advantage over his darker competitor. But in a sale ring the bright chesnut would command most money and most bidders. Many of the best buyers would not ask the price of a colt who showed signs of turning dark in his coat. The Stud Book describes many different shades of chesnut. The most objectionable is the dull, mealy chesnut, fading off at the flanks, muzzle and lower extremities, to a dirty white, sugar-paper hue, indicative of a weakly constitution, and want of fire and stamina. The cherry-red of sherry, strong beer tint, denotes a remote bay strain, but there are not many of this description. The real red chesnut, a few degrees darker than the standard bright chesnut, is a favourite colour with many. Crisp's Red Champion and Cordy's Marquis were both beautifully red, and kept their colour in all seasons. The dark chesnuts vary in different times of year, but the authorised bright chesnut, with tail and mane to match, never alters, in spring, summer, autumn or winter. It is by far the most prevalent shade in the best studs, and there are large stables of Suffolks where this is the only colour allowed. That champion Suffolk, Mr. Alfred Smith's Wedgewood, a son of Mr. Manfred Biddell's grand Prince Charlie, is of the orthodox true colour, and till we get back to the tenth generation in the male line, till we meet with Plant's Dark Horse, a grandson of Brady's Briton, the " very dark chesnut with a bit off his dock," there is nothing in the line but bright chesnuts.

Crisp's old Highnecked Captain, a son of Liverpool Captain, the winner of the Royal Argicultural Society's prize when they went there in 1841, was almost lemon colour, and left some capital fillies of what is called a yellow chesnut tint.

The silver-haired chesnut, a mixture of white hairs in a bright chesnut coat, has been a marked feature in some of the most noted Suffolks. Badham's Chester Emperor inherited this from the sire of his dam, Catlin's White-faced Boxer. The celebrated Cupbearer 3rd had a good many white hairs in a coat inclined to be dark. But all Suffolk breeders reject a horse if these silver hairs are plentiful enough to amount to a roan.

The late Earl of Stradbroke, who had known the breed many years, and was an exhibitor of Suffolks at the county shows, held that the dark chesnut was the most hardy colour, and regarded it as a sign of a good constitution. Others speak of the white mane and tail as a characteristic of the original breed, but these are not fashionable, and, as before stated, nothing is so good as the bright chesnut, with mane and tail of the same colour.

The date of the institution of the Royal Agricultural Society of England marks the commencement of a wider field of improvement in all breeds of agricultural animals in the British Isles. Long previous to that the Bakewells, Collings, Ellmans and other pioneers in stock-breeding had made their names known. Such men stood out in bold relief among even the best farmers of their day. They were head and shoulders above other stock-breeders, and could be named and numbered by anyone connected with agricultural pursuits. The meetings of the Royal Agricultural Society drew their successors together, and humble imitators took their products as models to work on. In a few years the leaders among these improvers no longer had the field to themselves. The Suffolk Agricultural Association was started some eight years before the first meeting of the Royal Society of England was held at Oxford. What the National Society did for England the County Association did for the district where the Suffolk horse had his home. The Crisps, the Catlins, the Cottinghams provoked a rivalry on the show ground which brought numberless competitors into the field, and defects which the old men had passed over as of little consequence, or had altogether failed to detect, were brought to light, and if the general character was not greatly improved there was one point, and that a vastly important one, to which the public drew attention. The outsider, the foreigner, the breeder's best customer, began to ask for a sound horse as well as a good-looking one. The judges in the showyard took a less lenient view of the matter than did the guests after dinner who strolled into the paddocks and said complimentary things of all they saw. Many an animal which at home was supposed to carry all before him, and did so up to a certain point in the showyard, was rejected by the veterinary, and some competitor with less to be said on his behalf on other points was ultimately placed first on the judge's list of awards.

The element of soundness began to be an important item. Loud and long were the complaints against the veterinary. But the committee of the Suffolk Agricultural Association adopted a drastic crusade against the side bone nuisance; rejecting bad eyes, roaring, and every ill that could be catalogued under the head of hereditary disease. For years the Association has ruled that no prize in the horse classes shall be awarded till the judge's selection has passed a veterinary examination. Well would it be for all breeds of horses if the National Society would thus take the bull by the horns and put the foot down on disease in this way. Unfortunately—incomprehensibly—many think the Council extend their examination only to certain classes, leaving the point of unsoundness in other classes to be decided by the judges, calling

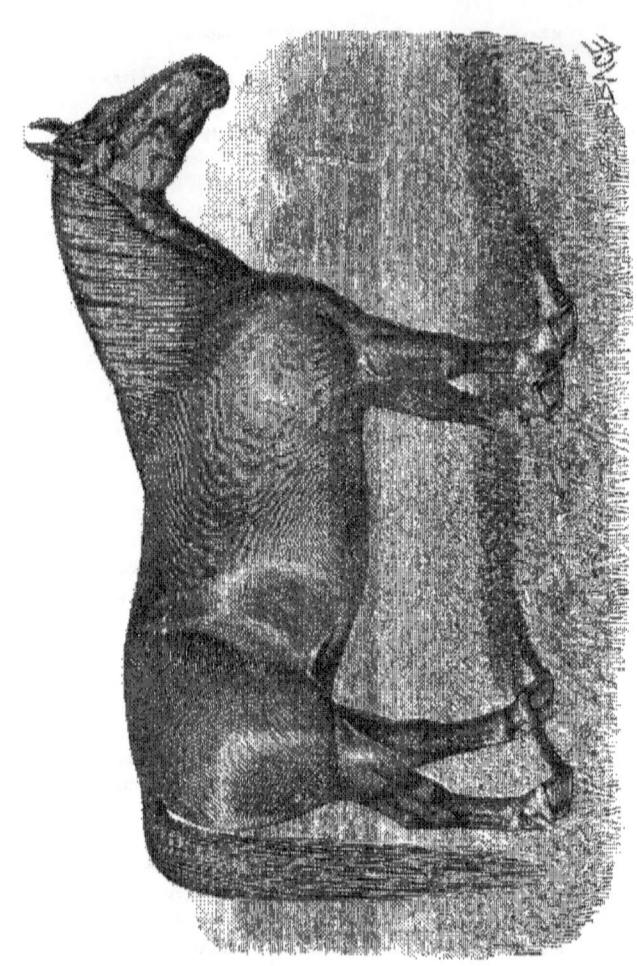

SUFFOLK MARE, BOUNCE 2517.
Winner of Queen's Gold Medal at R.A.S.E. Show, Windsor, 1889, &c.
The Property of Mr. William Byford.

in professional assistance only in such cases as they deem necessary. No judge should have this responsibility thrown on him, and the upshot is that animals which their owners dare not even enter for the county meeting have been sent to the Royal and have taken the first place in their class.

The effect of this salutary measure has been of the greatest service to the breeders of Suffolk horses. Thirty years ago the complaint of unsound feet and side bones was justly made in regard to Suffolk horses. It was this that stirred the County Association to combat the evil. The veterinary was chosen at the same time that the judges were appointed. The list of veterinaries who have acted for the Association comprises many of the very highest names in the profession. At a meeting some time ago one of the latter made the observation that although he had examined first, second and third in eight different classes, "he had not laid his hand on a side bone that day." As regards soundness, so far as feet and legs are concerned, the Suffolk at the present time stands at least on a level with the best breeds of draught horses in the United Kingdom. It was not always so. Those who believe that the defect is still there should take a look at the Newbourne Hall, Rendlesham, Playford, Glemsford Court and other large studs in the county. They will find no side bone curse there, and as good feet as the best stables in other counties can show.

As a horse for strictly agricultural work, as a farm horse pure and simple, there are few who will dispute the claim of the Suffolk to be among the very best in Great Britain. Active, of medium size, with perseverance at a dead pull, and capacity for a long day's work, he stands pre-eminent. The practice in Suffolk is very trying to a horse's constitution. From 6.30 to 3 o'clock is a long while to be at the collar, even with the half hour's rest (mostly without food or water) at eleven o'clock. There is no doubt the Suffolk stands this kind of treatment better than any other. In fact, it is doubtful if any other breed could stand it long at all. Whether the

Suffolk, there is no market for other than agricultural purposes in his own county.

It is, however, merely a matter of selection and attention to breeding. If horses of the pure Suffolk pedigree, with the substance and size of bone of Mr. Quilter's The Czar, Mr. Alfred Smith's Queen's Diadem, Foxhall (now in the United States), Prince Charlie (dead), and others that could easily be named; if sires of this calibre were put to picked mares of like size and substance, plenty of horses would be bred in the county of Suffolk which would command as ready a sale in London as the bays and browns found there. But all stock breeders of experience know that the attempt to breed outsized animals too often results in breeding disease. The pure Suffolk is not a large horse, but he is thick through him; his bone is much larger than his clean legs would lead a stranger to suppose, and his being so close to the ground makes him look smaller than he is. The old breeders in the county rightly prefer to keep to such a model. As long as America was ready to take his produce, while Australia and South Africa open their markets to him, he can afford to let the Shire-bred breeder have the markets which London, Liverpool, and the railway companies always present to the larger horses. There is room for both kinds—and both deserve all the encouragement they meet with—a fact it would be well if some who deprecate the chesnuts at the Hanover Square meetings would remember.

The old men in the county of Suffolk are apt to speak of a generation of horses dead half a century ago as being superior to those of our own time. Fifty years have probably made them more particular in their fancies; and if what their recollections place before them were here in the flesh, they would probably see faults and shortcomings long ago forgotten. In all probability there have been better horses of all breeds produced during the last twenty years than were ever produced before. It must be the case with Suffolks. These are sounder horses, better proportioned horses, better actioned horses. The type has been adhered to, but the veterinary's hard blows have not been without effect. In fact the Suffolk breeder owes much to the rough treatment the exhibits of five-and-twenty years ago underwent at the professional's hands.

One fact speaks volumes for the Suffolk horse. The best prices at the sales of the best animals are nearly always given by the county men. At the sale of the Butley Abbey horses in October, 1891, many were bought for distant counties and for Ireland, but it was the Suffolk breeders who gave the highest prices and made the average for mares a high one, and who made the twenty-nine foals reach an average of forty-four guineas each.

The winners of the principal prizes at the shows of 1892 sprang from four different branches of the old breed. The chart given herewith shows lines emanating from Smith's Horse of Parham. Through his direct descendants, Edwards' Old Briton 490, Crisp's Fairhead's Boxer 405, Catlin's Duke 296, and Newcastle Captain 89, we have four distinct tribes, three of the heads of which will be remembered by many of our breeders of the present day. There are some yet in the county who remember Edwards' Old Briton, but not many. Of this horse the late Mr. Barthropp, of Cretingham Rookery, used to say, "He was the best cart stallion I ever saw." What the heads of our veterinary departments would have said to him we know not. Some magnificent descendants of this horse had the side bone curse, but of the produce of the two branches by which he is now represented, no complaint in that way is ever heard. The nine generations between Mr. Edgar's first prize foal (Tittle Tattle by Flatt's Wantesden Duke) and Mr. Edwards' horse have little showyard honours to their credit. Mr. Flatt's horse was in the county many years, and was occasionally seen at the local shows, but he never quite reached a prize-winning standard. He had much of the true Suffolk character, but there was a lack of power in his limbs and joints. He was a beautiful chesnut, with a roomy car-

case, short legs and iron constitution. This made him a
favourite with many of the breeders in High Suffolk. The
Wantesden Duke strain has no Cupbearer blood in it, but
Mr. Edgar's Prattle, the dam of Tittle Tattle, the Bury prize
foal, is a daughter of Cupbearer III., and the infusion of that
popular blood may bring the stock of Edwards' Briton to the
front again. There is room for a good horse of this tribe,
but unfortunately Mr. Edgar's colt will be no cross for the
numerous Cupbearers, inasmuch as that blood is already
there. Unless Mr. Edgar's colt grows into a horse good
enough to be kept in the county, or his sire, Rattle, leaves a
popular son, in all probability the Edwards' Briton line, the
Julian's Boxer branch of the old breed, will become extinct
in the male line.

In the year 1879, the champion prize for the best Suffolk
horse at the Kilburn meeting of the Royal Agricultural Society
of England was won by another descendant of Edwards'
Briton. This was Mr. Manfred Biddell's Jingo, a passingly
handsome two-year-old colt. He had won at the home shows
the only times he was shown—at Woodbridge and Lowestoft.
Unfortunately his owner was tempted to sell him at the Kilburn meeting, and for years he was in Surrey. There he had
no chance, and to make matters worse, bad care, bad feeding,
or ill luck, sent him all wrong. A few years ago he was
bought out of a coal cart, a wreck of his former self, and he
was tried on the old stock at home. He left a few beautiful
fillies, and was afterwards bought by the Duke of Grafton;
but he rapidly became too lame even to get out of his box, and
he was shot, without leaving a son to perpetuate what some
good judges thought was the handsomest Suffolk ever sent
out from that stable at Playford from which so many good
ones have emanated.*

---

* Since the above was written Mr. Edgar's colt won the first prize for two-year-olds at the Woodbridge spring show in 1894. But the champion prize went to Mr. Pratt's Eclipse—yet another Cupbearer III.

Of this branch of Edward's old stock there is yet another horse in the county, or, at least, there was a year or two back. He, too, was bred at Playford, and was a half-brother to Jingo. For year after year, under the colours of the veteran Mr. Wilson, of Baylham Hall, Wilson's Vanguard travelled the Eye and Framlingham district. Year after year he left winning foals in the local shows of that part of Suffolk; possibly there may yet be a remnant of this branch to perpetuate the stock now so nearly extinct. A daughter of this horse appears in the chart as Lady 3035—Mr. Byford's three-year-old filly, which, although passed over at Bury St. Edmunds, stood first in her class at the show of the Essex Agricultural Society held at Harlow in July, 1892.

Wolton's Royalty 1339, a horse of which there is an excellent portrait by Duvall in the first volume of the Suffolk Stud Book, was of the strain of Edward's Briton. He travelled many years in Suffolk, left quite a number of winners, and was afterwards sent to Ireland. There may yet be a Royalty horse in the Newbourne Hall stables, for that fine old homestead has a large collection of Suffolks. If so, we may see a revival of the Julian's Boxer tribe through his son Edward's Old Briton, for Royalty comes in as a son of Magnum Bonum on the extreme left of the chart.

This Edward's Old Briton was a son of Julian's Boxer, one of the two branches through which Smith's Horse of Parham brought down the old breed in the direct male line. The notes in the Suffolk Stud Book under Julian's Boxer are very explicit. As the founder of a tribe for many years in the ascendant, and after figuring prominently nearly ninety years is still extant, we give the extract. Julian's Boxer was—" A bung-tailed horse, one of the four short-tailed sires his owner was in the habit of driving in a team together—'a nice upright red chestnut, with a white star.' The four were all advertised in 1815 [the other three were Briton, Bumper, and Bly] to travel the country 'till July 1st, if weather permit.' By all accounts he seems to have been an excellent system is a commendable one is open to question. He certainly works from year's end to year's end on rations that would reduce to a skeleton the larger specimens of the Shire-bred found at the Royal Society's meetings.

It was formerly said that a pair of Suffolks could plough more land (mixed soil probably preferred) than any other breed. They used to do so. Much of this kind of thing depends more on what follows than on what leads the plough. A horse's pace greatly depends on the man behind. Although no horseman can get through a great day's work with a pair of lazy drones, the most active, the best walkers, readily lose their pace if the ploughman is allowed to tie both, or even one, back to the whippletree. There is far too much of this going on in the county of Suffolk, and if the farmers there do not take the matter up the Suffolk horse will not long maintain his character as a plough horse.

A great deal has been said about the Suffolk horse not being adapted for the stones of the London streets. Another cry is that the Suffolk would be more saleable if he had a brown coat on. Both these assertions are open to question. Considering that the Shire-bred and Clydesdale supply to the metropolis is the gathering from Carlisle to Southampton, and from North Lincolnshire to the Land's End, and that the field for Suffolk horses is limited to three or four counties, the proportion of Suffolks seen in London, is far larger than is commonly supposed. But granting that the latter are not numerously represented in London, the reason has nothing whatever to do with his feet, or his legs either. In standing wear and tear they are a match for the rough-legged breeds anywhere in the world. The Suffolk horse is not of weight enough for London. As a rule he is not big enough. For the lighter vans nothing is better, but he has not power enough, he is not built on a scale heavy enough, for the stiff work of starting a two-ton load for the Manchester warehouse or the London dock. When he is found of massive build to compare with the Shire-bred, London work suits him well

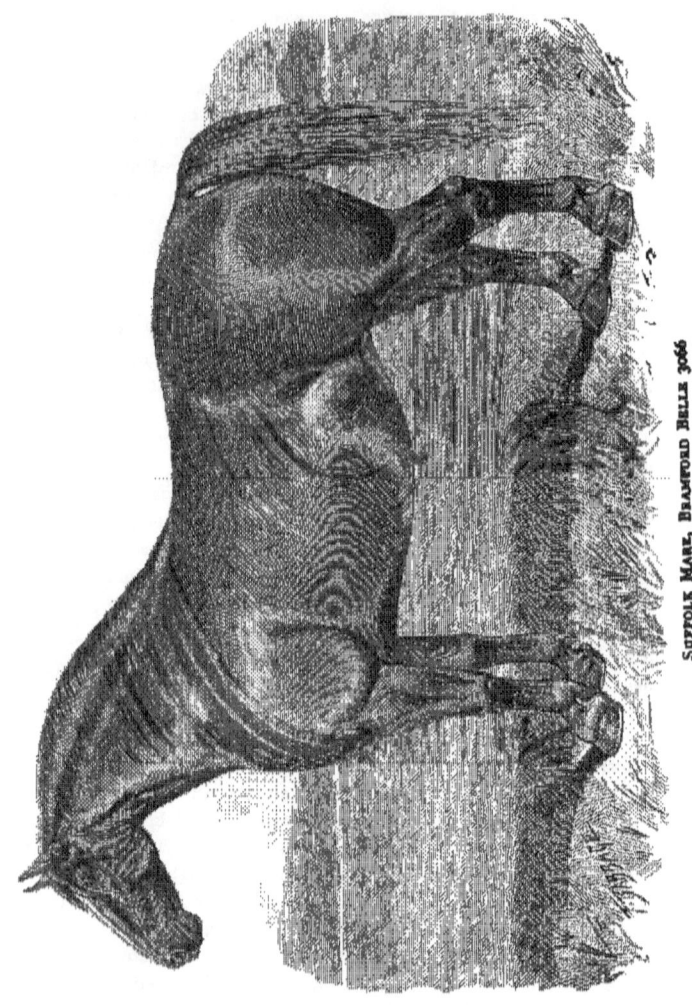

Suffolk Mare, Bradfield Belle 3066
Winner of First and Champion Prizes.
The Property of Mr. N. Catchpole.

enough; but if to breed for the market, regardless of his worth as an agricultural horse, is the object of the farmer, probably he will get his best return from the Shire-bred mare. As for the feet and legs not standing town work, nothing can be wider of the truth. For eight years a pair of good, roomy Suffolk geldings trotted side by side in a large railway van in one of the towns in the eastern counties. They were out early and in late; they were always in condition, and the work they did would have sorely tried those mammoths shown at Islington Hall in the spring; their tremendous knee action would have worn them up in half the time the chesnuts were on the stones in the town referred to.

It is said that the chesnut colour militates against the sale of a cart horse. It is a curious circumstance that those in Suffolk who have their best and largest geldings to dispose of never complain of this being the case. Not many months back a farmer sent a large Suffolk gelding to a local fair. He made eighty guineas of him. Before twenty-four hours he was sold for a hundred. The Suffolk breeder will not readily believe if he had been a bay he would have made more money.

Some years back, out of curiosity, the writer of this article went into the neighbourhood of London to see a stable of contractor's horses sold by auction. They were all Suffolks, out-sized specimens certainly, but they were pure-bred ones. They were just from hard work. They made enormous prices, and on inquiry they were found to be nearly all bought by dealers. One came from a farm near Ipswich, and was well known in that neighbourhood. The horse was eleven years old. He made ninety guineas. Others were sold at corresponding figures. Probably there are few Shire-bred geldings at eleven years old that would have made more. There is no doubt but these were exceptionally large, full boned, sound horses. It was, however, a proof that, if of size and substance enough, there is as ready a sale for good light chesnuts as for bays and browns. It must, however, be admitted that for the under-limbed, over-topped, neat little horse, and of what is still understood as the Suffolk type. He lived till he was close upon thirty, and travelled twenty-five seasons. One of his old leaders is still alive (1880), and his description of the horse is quite in keeping with that of others who have given the editor their personal recollections. He left some fifteen sons, all of more or less public repute, but it is through Edward's Old Briton 490 that the blood is represented at the present day. Julian's Boxer 755 was bought by David Wight, of Barningham (near Thetford), and travelled several years as Wight's Boxer 1388. Much of the best blood in West Suffolk comes through him and his son, a younger Boxer 1389, which Mr. Wight travelled in that part of the county. In Julian's Boxer the original strain of Crisp's Old Horse of Ufford 404 was united with direct descendant of Blake's Old Farmer 174."

Of this tribe, too, came the stock of Captains, one of which took the first prize at the Liverpool meeting of the Royal Agricultural Society of England, held there in 1841. Crisp's Conqueror, the sire of the first of the Cupbearers, came of a mare by a Crisp's High-necked Captain—the yellow, lemon-coloured horse, so well known for many years in East Suffolk, who was also a son of the Liverpool prize horse. This tribe is now extinct in the male line.

The Prince Imperials (a charming horse, light in his middle, but with round ribs, and extremely handsome), Wilson's Bismarck (a very large horse), Byford's Volunteer, Wolton's Monarch, Garrett's Viceroy, and Wilson's Heir Apparent, were all from the Liverpool Captain branch, but it does not appear that there are any descendants of these horses left at the stud.

We must now go back to Brady's Briton, the dark chesnut son of Smith's Horse. It is from this branch that by far the most numerous of the Suffolks of to-day take their origin. Turning to the extreme right of the chart, there are eight generations in the male line without any branches till we get to the sons of Mr. Biddell's Prince Charlie. Through one of these we get what is now apparently one of two lines which bid fair to monopolise the fashionable strains in the showyards. In strong contrast to the line on the left of the chart, we have here link after link in the chain well known at home and at the Royal Agricultural Society's meetings as first-prize horses. Of the horses on the left of the chart, Bush's Albert, Partridge's Albert, Allen's Albert, and H. Biddell's Champion, not one was a winner at home or abroad, and, with the exception of the latter, not one even saw the inside of a showyard. On the other side, since the days that agricultural shows were instituted, from that grand filly, Dainty Dolly, now in Mr. Henry Showell's stud at Playford, to the eighth generation backward, there is but one name which is not found in the catalogues of the Royal Agricultural Society, and the Suffolk Agricultural Association. Plant's Captain was exhibited at the second meeting of the Royal, held at Cambridge in 1840, and although the winner turned up in a Lincolnshire horse, Captain was sold into Essex, and when twenty years old again changed hands, but still remained in that county. He was first at Saxmundham in 1837, and second a year before at Wickham Market. A note in the Stud Book says "He had his elbows turned in; a bright chesnut, with short legs, and was a good horse of his time." He was foaled more than sixty years ago. The next generation was a noted horse for many years in Suffolk. This was the late Mr. Barthropp's favourite old Newcastle Captain, with which he won the £40 which the Royal gave as first prize among the cart horses (all breeds) when they visited the north in 1846. He was rather a small horse, somewhat light in his body, but round in his ribs, with a fine upstanding foreend and magnificent hind-quarters. He, too, was a little confined in front, a fault he transmitted to his son bred at Thurlston, near Ipswich, when the late Mr. Badham lived at the Sparrow's Nest. He was first at the meeting of the Royal Society in 1858, and was afterwards known as Chester Emperor. There is a portrait of this horse in the Suffolk Stud

Book, taken from a photograph. He had no more hair on his legs than Blair Athole, but they stood the wear and tear of a constant round on the hard roads for years, and at the time of his death, after carrying an immense mass of fat from year's end to year's end, those legs were as clean and puff-less as they were when he was a two year old.

His son, Harwich Emperor, was as fine a horse as ever was bred by anyone in any county, but his forefeet and ankles were faulty. He, too, was a Royal winner, and for a time carried everything before him in his own county. He was first at Oxford, when the Royal Agricultural Society paid their second visit there in 1870, and was then bought by the Stonetrough Colliery Company, and under that ownership he appeared as "reserved number" at the Wolverhampton meeting a year or two afterwards. A son of this horse was bred at Creeting, near Stowmarket, in the year 1871. This was Rainham's Prince 1002. His breeder was the late Mr. Maurice Mumford, from whose stables, first and last, many a good Suffolk was sent into the showyard, or sold to go abroad. But Rainham's Prince never figured before the judges, and was the solitary exception of the family in that respect. But he left a son who, had he lived, would probably have made up for his sire's obscurity.

Mr. Manfred Biddell's Prince Charlie was a grand horse indeed. He took first prize at Shrewsbury and first prize the next year at the Woodbridge Spring Show, but died the month following. He lived long enough, however, to leave two excellent sons; one of them was Mr. Everitt's Warrior, well known in the show yards of the present day, and the other Mr. Alfred Smith's Wedgewood—a horse that since he was a two year old has never been beaten. It is sufficient to say that at the summer shows although his oldest colts were only three years old, they numbered eight among the winners at the Royal and County exhibitions of the year 1892. A photograph of this most charming horse will be found in the sixth volume of the Suffolk Stud Book.

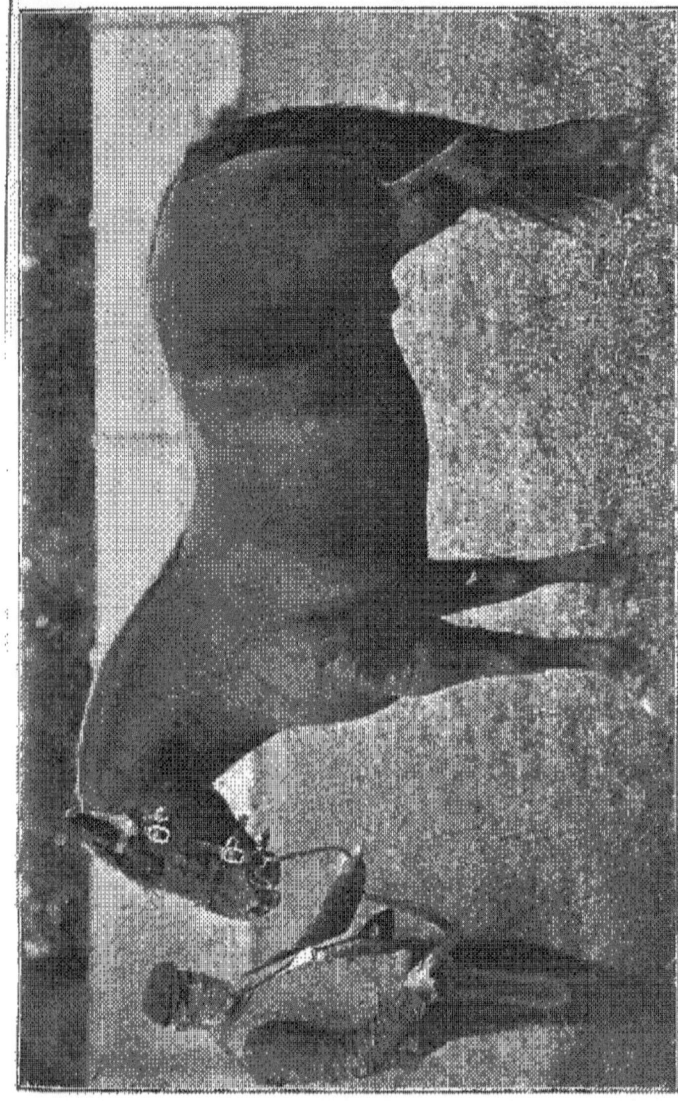

From a Photograph by H. R. Gibbs.]

SUFFOLK STALLION, PRINCE WEDGWOOD 2364.
Winner of First Prize at R.A.S.E. Show at Darlington, and at the Suffolk County Show, 1895.
The Property of Mr. W. Cuthbert Quilter, M.P.

The third line by which the stock of Smith's Horse comes down to us is through Groom's Ramper, another son of Brady's Briton. At the fifth generation the stream divides and we have the two sons of Manchester Boxer—Catlin's Duke and Crisp's Fairhead's Boxer. At one time the former was by far the most popular horse of his day. In proof of this the large Stud Book chart shows forty-one of his sons, all of which appeared in the showyard catalogues of their day, or were at the stud. He was a great prize winner. He was first among the aged horses at the Windsor meeting of the Royal Agricultural Society held there in 1851. He was a very handsome, bright chesnut, with a snip on his nose and a small star on his forehead. His back was cloven all the way along, with the muscles lying up side by side like a ram. With great girth, round muscular quarters, and a fine temper, he was a vast favourite with those who clung to the real Suffolk type; but he was a little small in the second thighs, and although his hocks were sound and bony they were not large. He was bred at Butley Abbey, and at the death of Mr. Catlin was sold for 260 guineas. He was then far beyond his prime, but he was in Essex some years, and died in the possession of Mr. Fisher Hobbs. His sons and grandsons were legion, but the showyard celebrities of the present day come through only two of them—Wolton's Champion and Capon's Duke.

From Capon's Duke through seven generations of popular horses—showyard winners for the most part—we get Wilson's Old Times, a thick-set, dark chesnut, of genuine Suffolk build, but of a grade just missing the first rank. His young stock are taking first places all along the line. Foremost among these is Mr. Wilson's Suffolk Model, bred at Baylham Hall. He was first in a class of thirteen capital yearlings at Bury, and, moreover, looks like growing into a future winner. Memory is a superb yearling filly; she, too, was bred at Baylham, and was sold as a foal at the Framlingham Show to the late Mr. Hume Webster. At the Marden Park sale the Duke of Hamilton's agent bought her for his Grace, and although the price was a long one, there was nothing to regret in the purchase. The Old Times foal out of Bramford Lass was third in a large class, and so altogether Mr. Wilson has good cause to be satisfied with the future prospects of Old Times as a Suffolk stud horse. Here too we may note the oft-repeated circumstance that the blood of a noted horse is carried on through generations when apparently the line looks like dying out, and then a sire comes into notice and once more the strain takes a foremost place. Catlin's Duke may yet be the favourite blood in Suffolk, and if Mr. Wilson's horse gets a good season or two through his yearling victories, in all probability, it will be so.

The next branch of the chart we have to notice is the other son of Catlin's Duke. In the year 1850 the prize foal at the Suffolk Society's show, held in Christ Church Park, Ipswich, was a tremendous out-sized red chesnut, bred at Newbourne Hall. He grew up with much quality, but of a size too big, a hand too high, for East Suffolk. Mr. Wolton sent him into the west side of the county where he left some very good colts. One of these, bred at Wickham Brook, some ten or twelve miles south-west of Bury St. Edmunds, was a remarkably heavy horse, with great bone, but on much shorter legs than his sire. In his declining years he was bought by the late Mr. Crisp, and after being freely used in the Wickham Market district he was sold at the Butley Abbey sale to Mr. Catchpole, of Bramford, in whose possession he died. His best son was undoubtedly Captain Snap, a horse bred by Mr. Manfred Biddell, used some years at the Lux Farm, Playford, and then bought with six other Suffolks to be shipped across the Atlantic. Fortunately Captain Snap had a cold, and was kept at Glasgow for a later shipping. Of the other six, four were so knocked about by a fortnight's terrific weather that they died and were thrown overboard; two only of the half-dozen being landed alive. Captain Snap was heard of again, and was always referred to as a great catch for the cart horse breeders in America. But Mr. Biddell had a colt in his yards at the time he sold the old horse destined to take a great part in the show-ring and at the stud. There was Ben —Biddell's Ben, as he was termed—a very splendid specimen of the Suffolk breed. A beautiful portrait of this horse, by Duvall of Ipswich, adorns the dining room at the Lux Farmhouse, and from which a capital lithograph was taken for one of the illustrations of the first volume of the Suffolk Stud Book. He was a large horse, as handsome and full of quality as he was big, with a bone below the knee of something considerably over eleven inches. He was good enough to win the Hundred Guinea Challenge Cup at the Ipswich meeting of the Suffolk Agricultural Society in the year 1878. He was sold to Mr. Kent, and ended his days somewhere by the banks of the Thames.

But there was yet another generation of the same family, and he too was bred at Playford. This horse was sold as a yearling to Mr. Hurrell, of Newton, near Cambridge. Here he stood year after year. At last that astute business man, who does so much for Suffolks in the Cambridge and Essex corner of the county—Mr. Byford, of Glemsford—took a fancy to him, put him in show order, won a prize with him at Lowestoft and another at the Royal at Windsor, and then sold him to go to Sweden. And Mr. Byford bought a son of Playford, bred by Mr. Hurrell, out of a mare also of his own breeding. This was the massive colt Surprise, a Playford-bred horse on both sides, for his sire was foaled at the Hill Farm, and the sire of his dam, Biddell's Champion, was the stud horse there for several years. Surprise has all the muscular frame and short legs of that long line of heavy Suffolk horses, reaching back to the old Wickham Brook horse of Mr. Bromley's. He was first among the aged horses at Bury. Here is another chance for Mr. Catlin's Duke strain, for those who aim at breeding horses of size enough for town work will no doubt patronise Mr. Byford's Bury prize horse.

But Ben had yet another son, like Playford very dark in his coat (a shade strong in the pedigree of Ben's dam, for she was a daughter of Law's Farmer, quite a black chestnut). This was Catchpole's Champion—a short-legged, thick-set horse, good enough to gain the first prize in his class at the Suffolk show at Ipswich in 1884, as well as the Waveney Cup as best horse in the yard. As he is, or was till lately, still in good preservation, there may yet be another line of Catlin's Duke horses to remind those who remember him of the old Butley Abbey favourite. From Mr. Catchpole's he was transferred to the Rendlesham stud, where he had a fine chance, and was then passed on to Mr. Lewin, of Dunningworth Hall.

We have one more line to follow in the chart. In 1846, the then Duke of Manchester bought of old Mr. Catlin a wide-eared, finely built son of old White-faced Boxer. This White-faced Boxer was perhaps the last horse of note which bore strong likeness to the old breed described by Young. He was rather a small horse, low in the back, very short in the leg: had a large carcase, and was bent in the hocks. Manchester Boxer was larger, a grander horse to look at, but his ears hung out in so unsightly a fashion that he had to have a bridle made on purpose to keep them in position. Besides being the sire of Catlin's Duke, he left another son which became famous in his day, although he never won a prize. This was Crisp's Fairhead's Boxer, so called to distinguish him from the numerous race of that name. He was bred by a miller, three miles from Woodbridge, who had a small farm as well as his mill. He passed into other hands, and after being located for some time in the district where the sires from Newbourne Hall, Kesgram Bell Farm, the Playford stable, and John Lewis' little place at Foxhall, got all the best mares, he was bought by the late Mr. Crisp. He was not a handsome horse, and up to his last change of owners had never been seen in show trim. But Mr. Crisp gave £100 for him. When his time came, and he had some of the best of the mares in the Crisp district, the produce fully bore out Mr. Crisp's judgment in buying him. He was a particularly square-made horse, but his hocks were bent, his arms were small, and there was a want of muscle on his back; but he was a short-legged, wide horse of beautiful quality, and with the temper of a pet lamb.

Among other capital horses owning him as their sire was an immensely heavy horse, but with bent hocks, if possible more noticeable than his father's. This was Crisp's Conqueror, the son of a charming mare bred by one of the Messrs. Toller, since so famous among the Suffolk breeders. He was sold to the Prussians—then, perhaps, the best customers for Suffolks—at three years old. But he left a son, bred at Marlesford, destined to be the first of three generations of one name, all of the highest character and prestige. This colt was bought as a foal by Mr. Crisp, and never was sanguine hope more fully realised than the purchase proved. This was "old" Cupbearer—Crisp's Cupbearer—a copper-coloured chestnut, winning local, county, and Royal prizes till long after Mr. Crisp's death. He had immense girth behind the shoulders, great width in front, and had a wonderfully free and bold action in the ring, but he too, was bent behind, was a little light in the flank, and as a noted all-round judge of horses—long since "gone over" to the Shire-breds—once remarked, "He had quite as many faults as a good one ought to have." Suffice it to say he won first prizes at home time after time, won the three-year-old prize when the Royal Society came into Suffolk in 1867, and after Mr. Crisp's death was sold by auction to Mr. Richard Garrett for 370 guineas. In his hands he won the first prize at the Royal when the show was at Wolverhampton, the first at Beccles the same year, and afterwards passed into the hands of a clergyman in Suffolk who took to stock-breeding, and died in his hands.

From this horse the late Mr. Frost bred Cupbearer 2nd, afterwards the property of Mr. Catchpole. He was a finer horse than his father, had fewer faults, and was a beautiful bright chesnut. Unfortunately, he died shortly after he was bought by Mr. Catchpole. He had, however, won prizes enough to make his mark in the showyard. He was first at the Bedford meeting of the Royal Agricultural Society, and first at the County Show at Stowmarket in 1875. He left two sons, Garrett's Cupbearer the 3rd and Wolton's Chieftain. The former was, perhaps, even better known in the showyard than either his sire or his grand-sire. He was bred by Mr. Frost, the breeder of the second Cupbearer, and was sold as a yearling at a sale of his colts at Ipswich for 200 guineas. Mr. Garrett never regretted the purchase, for his winnings in prizes amounted to nearly three times that amount. He was a remarkably heavy horse, very short-legged, wide almost to a fault, and not very bright in his colour. He had the bent hocks of his ancestors, a weakness from which his sire was free. At Mr. Garrett's death he was again sold by auction, and made double the price he cost as a yearling. As a stud horse, judged by the number of winning colts and fillies he left, he was almost unequalled, and, singular enough, almost to the last he left as good stock as ever. Mr. Austin, of Brandeston, gave 400 guineas for him, and he afterwards passed into the hands of Messrs. Pratt, in whose possession he died about three years ago. A glance at the chart will show even at the most recent shows how strongly he was represented in the prize winners. At Bury St. Edmunds in 1892, Pratt's Eclipse was first as a three-year-old, and was awarded the cup as best horse in the yard; the Duke of Hamilton's Queen of Hearts was first as a brood mare; his Grace's Queen of Trumps was first as a two-year-old, and his Morella won second honours in the gast mare's class. Mr. Edgar's Prattle was second in the brood mare class, and these were all sons and daughters of Cupbearer 3rd. Among his grandsons at the same show, Messrs. Pratt's Earl was first among the two-year-old colts, and Mr. A. Smith's Democrat was a prize winner at Warwick.

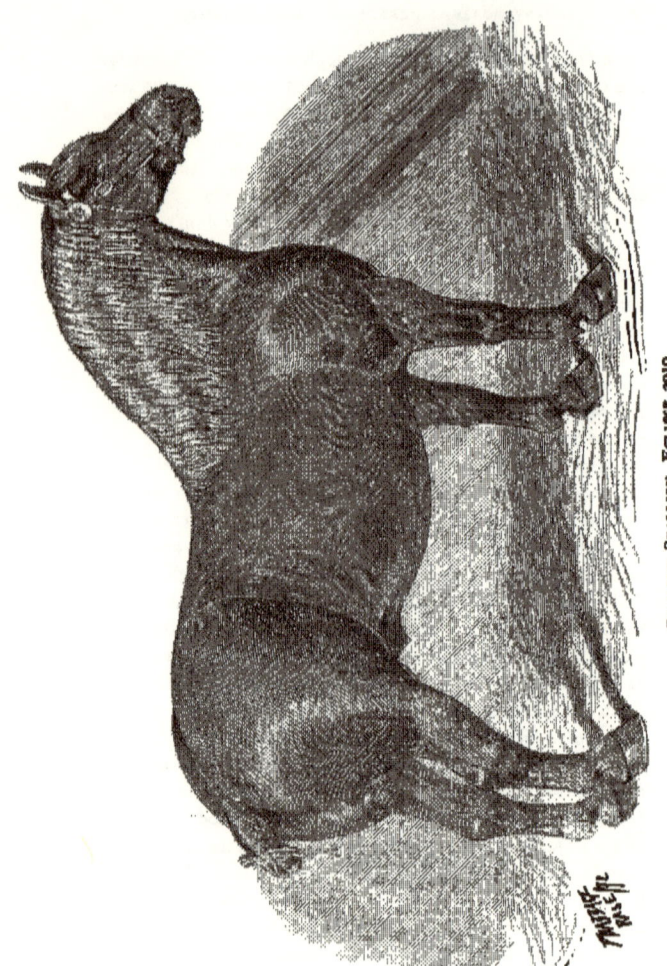

Suffolk Stallion Eclipse 2010
Winner of First and Champion Prizes, 1891, 1892 and 1893
The Property of Messrs. T. Pratt & Son.

Through a son of Cupbearer 3rd, Rodney, we get that magnificent horse, Mr. Biddell's Foxhall, a Royal winner at Shrewsbury, and a frequent winner at the county shows. He was a rare stamp of horse with the large bone, muscular frame, and fine action which take with the Americans, and to America he went, under the colours of Mr. Galbraith. He came of a mare by Captain Snap, her dam by Crisp's Old Cupbearer, a double strain of a popular tribe, and he fully bore out her breeding.

A son of Foxhall, another of Mr. Biddell's, won first prize at the Woodbridge Show in March, 1892, as neat a specimen of a thick-set, medium-sized Suffolk as can be found in a day's drive. He had a great season that spring, and was only placed third at Bury in the summer.

From Cupbearer 2nd there is another branch of the strain in Wolton's Chieftain. The first prize brood mare at Bury, the second prize three-years-old colt, the second prize two-years-old colt and the sire of the third-prize foal are all sons of the handsome Chieftain. Few more beautiful horses than this have been led into a show ring. He was first at the Royal at York in 1888 and has won much in the home county also. He was a little taller than the Suffolk breeder thinks right. He was light in the hind-quarters, and had too much daylight below, but he was a noble horse in good condition, and has left some superb, bright chesnut, high-spirited mares. He died at Newbourne Hall some years ago.

This must conclude the sketch of the ancestry of the winners of 1892. To have taken in all the previous winners of the strains we have followed would have been beyond the space of a reasonable sketch, and the chart would have become an unmanageable sheet. It gives some idea of the origin and development of the tribes now in fashion, and may be useful to some for that purpose.

In Suffolk, as in other horse-breeding districts, the further restrictions on the imports of live stock to America have given a check to the exportation of prime animals from the county.

The Suffolk breeders have felt the effect of this, but the Australian colonies and South Africa are still open markets. The depression in the agriculture of East Anglia, the drop in rents, the serious reduction of the incomes of the landed proprietors who have no resources beyond their estates, have prevented many from keeping up their studs of Suffolk horses. The Duke of Hamilton has a splendid collection of Suffolk mares, and still buys when there is anything of marked value for sale. Mr. Quilter, the member for the Sudbury Division, takes good care to have a Suffolk horse for his district, and has some excellent mares of his own besides; but beyond this there is no great support given to the tenant farmers to encourage the breeding of the horses which have made the county famous, in so far, at least, as active participation in the pursuit is concerned. The county gentlemen do, however, subscribe liberally to the prize fund for Suffolk horses at the meetings of the Suffolk Agricultural Association, and in this way help to keep the breed before the public. But with the exception of the Duke of Hamilton already mentioned, there are no patrons of the breed to inflate the prices of the Suffolks, as many noblemen and other wealthy patrons do the Shire-breds. The agent of His Royal Highness the Prince of Wales makes no sign, and with the exception of a purchase made at the Kilburn show Her Majesty's name is never seen in connection with Suffolk horses. The exhibitors of Suffolks have an idea that they are treated with scant justice by the Royal Society. Certain it is that year after year the chesnuts are hid in the far, far away of the showyard; and but for the admirable representative the breeders have on the Council in the person of Mr. Alfred Smith, whose magnificent stud at Rendlesham is known all over the world, the overwhelming preponderance of Shire breeders on the Board of Management might forget the respect due to a class of horses which, till the breeds were separated, carried all before them year after year in the Royal Showyards.

Allusion has been made to the persistent adherence to

Suffolk Mare, Queen of Hearts 2116.
Winner of First Prize at R.A.S.E., 1890, &c.
The Property of the Duke of Hamilton and Brandon.

characteristic type which the breed of Suffolk horses has always asserted, in spite of the repeated attempts to engraft outside blood on the parent stock. Many a failure can be traced to disregard of the rule Nature in this way has so emphatically pronounced. She insists that the breed shall retain the salient points which marked the original stock. They were never large. Mark the result of those whose efforts have been directed to the attainment of more size. That is easily obtained by selecting for use out-sized stallions. But thence come disease, roaring, and the over-topped legs; perchance a grand animal offhand, but one which sooner or later the judge or the veterinary most surely rejects—a type, it may be added, which has done more to prejudice the breed than any other.

The bone of the Suffolk horse looks small, but denuded of hair and skin, more is left than on many a rough-legged rival. Those who, years ago, attempted to introduce what they took to be more substance, invariably failed, for with the big bone comes what is too often mistaken for it, the thick skin, the coarse hair, and the disease which seems indigenous to the lower extremities of the larger breeds. Marked instances of this will occur to those who watched the efforts of the breeders who, before the Stud Book was started, tried the effect of an outside cross. Failures in the opposite direction may occasionally be traced to in-and-in breeding from one strain of the pure breed; but by careful selection of the best stallions, breeders in the county of Suffolk now send into the showyard horses with feet and legs which no unprejudiced judge can find fault with.

No, the Suffolk horse is a short-legged, clean-boned animal, of ample size for any agricultural work in any district in England, and admirably fitted for active town work as well. He should be deep in the carcase, wide in front, square behind, with hard, short legs, close-knitted joints and devoid of all tendency to coarseness. Unless extremely well put together anything over 16·1 should be viewed with suspicion. If those who make trial of the breed will keep such a type in their eye, the significant signs of antiquity of origin will take care of themselves. The chesnut colour, the marked capacity of thriving on a scanty diet, and long hours in the collar will be there. And so will the docile temper, the never-ending patience at the dead pull, and the many days so remarkable in the age of the Suffolk horse.

Fortunately the Stud Book has saved the disappointment which the introduction of the out-sized cross has so repeatedly inflicted on the experimental breeder; but the hints here given may serve to warn those who, in starting a stud, imagine that by selecting flashy seventeen-hand specimens to breed from, they are going to produce a more powerful animal suited for town work.

CHAPTER III.

THE CLYDESDALE HORSE.

So much has within the past twelve or fourteen years been written regarding the history of the Clydesdale breed, that there does not appear to be much room left for anyone to advance new theories or unearth unknown facts on the subject. The records that have been examined and the views that have been advanced by various writers have generally gone to support the popular conception—that the breed is a composite one, and that the first recorded element in its composition was the use of Flemish stallions on the native mares of Lanarkshire about the close of the seventeenth, and in the first quarter of the eighteenth, centuries. There is, however, reason to believe that Flemish stallions had been introduced into Scotland long before the date mentioned, and records of an earlier period show that Scotland was recognised as a breeding district for horses during the early Stuart reigns. There was a trade in horses between Scotland and the Continent of Europe in those days, in which the Douglases—the ancestors of the Hamilton Ducal line—played an important part; and so extensive was the trade, that while in the reign of King James I., the Poet King, in the fifteenth century, all horses over three years old, were permitted to be sold for exportation, during the Regency of the Earl of Moray in 1567 an Act was passed prohibiting exportation. During the intervening century and a half great efforts had been made to improve the breed, but we can hardly think that these could have had much uniformity of success, especially as it would appear from the records, the instrument of improvement varied according to the tastes of the reigning monarch. At one time a horse capable of bearing heavy armour was the object aimed at, at another something very like what we would call a draught horse, and at a third a horse whose leading qualification was speed. Sir Walter Scott must be quoted as a believer in the view that the Flemish horse was in use in Scotland at a very early period, because in the "Fair Maid of Perth," chap. viii., he represents the gallant smith as riding "on a strong black horse of the old Galloway breed," and the honest bonnet-maker as perched upon "a great trampling Flemish mare, with a nose turned up in the air like a camel, a huge piece of hair at each foot, and every hoof full as large in circumference as a frying-pan."

Of course, the popular novel is not history, and Sir Walter's opinion may be of small value in such a case, but the historical setting of his writings is so generally conceded to be in the main accurate, that the probabilities are that he had good grounds for the opinion he plainly held. However, the class of horse, in use in Scotland before the Revolution settlement of 1690 gave the land rest, is not of any more than antiquarian interest, because the horse bred after 1690 would be required for a very different purpose than that of conflict and war. And therefore, we may safely conclude that the eighteenth century importation of Flemish stallions, if it can be established, has a much more important bearing on our present inquiry. Indeed, it seems to us that the history of draught horse-breeding in Scotland may be divided in three sections: the traditional, ending with the Revolution settlement; the historical, confined mainly to the eighteenth century, and having its records in the Old Statistical Account prepared under the superintendence of Sir John Sinclair, Bart., and completed in 1810; and the systematic, embracing the period from that date until now. We have now exhausted all that need for the present purpose be said regarding the first, and in connection with the second the question to be considered is, the introduction of the foreign influence which, conjoined with the conditions of agriculture in Lanarkshire, produced the modern breed of Clydesdale or Lanarkshire horses.

The introduction of these foreign influences are variously credited to the Duke of Hamilton of the period, and a farmer named John Paterson, of Lochlyoch, in the Upper Ward of Lanarkshire. The importation made by the Duke of Hamilton, is placed about the middle of the seventeenth century, and the place where tradition locates his six fine black "Stallions from Flanders" is Strathaven Castle. This theory or tradition is adversely criticised by Aiton, a Lanarkshire lawyer, who wrote much on the Agriculture of some of the Western Counties of Scotland, about the beginning of this century; but it is accepted by the writers of several of the Statistical Accounts. A modified form of the tradition, and probably the fact which forms its kernel, was held and regarded as an established fact by the late Lawrence Drew, tenant of the farm of Merryton Hamilton, who died in March, 1884. His theory was that James, the sixth Duke of Hamilton (1742-1758) imported a Flemish stallion, dark brown in colour, which he kept for the benefit of his tenantry, who were granted its use free of charge. The grounds on which this theory was held by Mr. Drew are such as will be regarded as satisfactory in dealing with historical data, and it may, therefore be concluded that such a Flemish stallion was in use in Lanarkshire about the middle of the eighteenth century. Another tradition of a similar character, and equally well authenticated, is that to which the compiler of the Introductory Essay to the Clydesdale Stud Book (1878) is committed. This is the tradition which connects the introduction of the Flemish stallion with John Paterson, of Lochlyoch, in the parish of Carmichael, about the years 1715-1720. It is undoubted that the Lochlyoch mares had a special reputation during the latter half of the last and the first quarter of the present century; and the family tradition is all the more credible because it was a proved custom of the Paterson family to chronicle important events in their history. As late as the year 1836, an Edinburgh newspaper in reporting the doings of a day's ploughing given to one of the Patersons, who removed from Lochlyoch to Drumalbin, refers to the fact that it was a brother of the grandfather of the new tenant of Drumalbin who brought the notable stallion from England to Lanarkshire, which had founded the famous Clydesdale breed of horses. We may, therefore, at least conclude, in view of these various traditions, that the Flemish stallion had something to do with the moulding of the type of draught horse known as the Clydesdale.

But while admitting to the full all that may be urged in supporting this view we are of opinion that there are certain circumstances connected with the development of the Clydesdale breed which have not received the attention which their importance merits. What first strikes one in an historical inquiry about the breed is its name. Why Clydesdale, and not Scottish? As is generally well-known, Clydesdale is the old name for the valley of the River Clyde, or in other words, for the most fertile portions of the great county of Lanark. "Paisley shawls" and "Kilmarnock bonnets" tell at once that the particular patterns of shawls and bonnets referred to are, or have been, in some particular way identified with the great industrial centres whose names they bear. Clydesdale horses, in the same way, must be regarded as horses, the development of whose type and qualities must, in the first instance, be traced to the conditions of agriculture in the county of Lanark. In other words our theory is, that the new conditions of life to which the peaceable and industrious inhabitants of Avondale and Clydesdale were permitted to apply themselves after the close of the Covenanting struggle, enabled them to improve their stock of horses to such a degree that the horses of Lanarkshire were famous at an earlier period than were those of any other part of Scotland. The climate and the soil of the Avon and Clydes dales are admirably qualified for stock breeding and rearing; and while the Flemish stallions and the later reputed English stallion, Blaze, which came on the scene about the close of the century, undoubtedly contributed much to the development of the breed, their influence was greatly enhanced by the favourable character of the soil on which it was exerted. It is clear, from recent experience, that external influences alone will not ensure improvement in the horses of a district or county. Sires that are credited with most favourable results in one locality have but indifferently distinguished themselves in others; and it is not necessarily in the district in which agriculture is most advanced that the best results in horse-breeding are secured, but rather under somewhat more primitive conditions. A pastoral country is the most favourable for horse-breeding; and until the mineral deposits of Lanarkshire began to be developed few districts could have excelled many parts of it as pastoral lands. The supremacy of Lanarkshire, and at the same time the restricted area of horse-breeding in Scotland, are well illustrated in the Old Statistical Account from the summary of which, in Sir John Sinclair's account (1812), we make the following quotation; vol. 1, p. 143:—

"From the high prices of horses a number of farmers endeavour to rear on their own farms a considerable proportion of the stock they require, though in many districts they depend on the western counties of Scotland and the northern counties of England for a supply."

This refers chiefly to the great agricultural areas of the Lothians and Berwickshire. And at page 146 in the same connection there is the following interesting passage:—

"It has been very justly observed that farms dedicated to the sole purpose of breeding horses, would certainly pay well at present if the necessary attention were paid to the breed and management. Such farms are very much wanted as an important link in Scottish husbandry, for the breeding of horses in the west of Scotland will be always diminishing as the farmers become better acquainted with improved arable management. There might probably be suitable situations found in the northern districts for that purpose, especially if more winter food could be obtained by the cultivation of fiorin or of Swedish turnips."

In advancing to consider the next stage in Clydesdale breeding, it will be seen that the surmise of the writers as to the issue of improved methods of agriculture in Lanarkshire came to be fulfilled, and the horse-breeding area was to a very large extent transferred to the pastoral and turnip-growing counties in the south-west and north-east of Scotland.

BREEDING: EARLY SIRES.

The consideration of the progress of horse-breeding in Scotland, under the somewhat more systematic conditions of the nineteenth century, is greatly simplified by the work that has been done in connection with the compilation of the Clydesdale Stud Book. In fact, no line of treatment is possible in this connection which does not presuppose the existence of the pedigree record.

The earliest known Clydesdale head of a family is, of course, Glancer 335, generally known as " Thompson's Black Horse." It is significant that most of the best known modern Clydesdales trace their descent in at least one line, and some in more than one line, from this celebrated horse ; and if the theory of the writer of the Clydesdale Introductory Essay already referred to be correct, it is an easy matter thus to connect most of the leading families of Clydesdales in the present day with the old Lochlyoch breed, descended from the black stallion brought from England by John Paterson, about the time of the first Jacobite rising. The theory referred to briefly is: that the Lochlyoch family of Paterson's and the Shott's Hill Mill family of Clarkson's being related, and continual intercourse being kept up between them in trade, there is every reason to believe that the Lampits mare—the dam of Glancer 335— which was bought at the displenishing sale at Shott's Hill Mill, in 1808, was descended from the Lochlyoch breed.

From all that can now be learned, Glancer 335 seems to have been a horse with more than a merely local reputation. According to the entry in the Clydesdale Stud Book he was foaled about 1810, but we are strongly of opinion that this is too early a date. According to the declaration of John Carr, who claims to have exhibited Paton's Horse of Bankhead at the Highland and Agricultural Society's Show, at Edinburgh, in 1842, where he gained second prize, Thompson's Black Horse was the sire of Paton's horse, and the latter was six years old at the date of the Show. He would thus have been foaled in 1836, and his dam would have been served in 1835. John Carr knew and wrought his dam, and held her when she was mated to Glancer 335. If this horse then had been foaled in 1810 he would have been twenty-five years old at the date of this service, and there is hardly any probability that this would be the case. A travelling card for him is published in the Introduction to the second volume of the Clydesdale Stud Book, from which it appears that his terms were one guinea, and a shilling to the leader. Unfortunately there is no date on the handbill. The next epoch-making horse is more notable, or perhaps it would be more correct to say, that the records regarding him are more complete. This was Broomfield Champion 95. In spite of the comparative familiarity of his name there are very few mares amongst the prize-winners of his time that are claimed as his progeny. His reputation rests more on his having been the sire of one pre-eminent horse, Clyde *alias* Glancer 153, or Fulton's Ruptured Horse; and several of his female progeny have a reputation, if not for showing, at least for breeding. One of them was a Haughhead mare, the dam of Farmer 283. Another, and quite a celebrated one, was the Lumloch mare, dam of Farmer 284, the sire of Victor 892 and Salmon's Champion 787, while yet a third was dam of the noted Highland and Agricultural Society's first prize horse Grey Emperor 369. There was another well-known breeding mare got by him in the possession of Mr. Paterson, Waterlee, Houston. She was dam of the Waterlee Famous Horse 903, or Scotchman 749, and, we are disposed to think, granddam of Lofty 467. There still lives (June, 1892) one who had two colt foals got by Broomfield Champion 95, Mr. Archibald Bulloch, Milliken, New Kilpatrick, Dumbartonshire. These two colts were of some repute in their time, but from causes which need not be detailed, their identity is somewhat lost in the pedigree records. The omission will be best supplied by explaining first that Glancer 338 and Superior 836 are duplicate entries of one horse. The former is the more accurate entry of the two. He was bred by Mr. Duncan, Glendivine, Winchburgh, who sold him to Mr. Joseph Bulloch, now tenant of Low Leathes Farm, Aspatria. Mr. Bulloch sold him to the gentleman named as owner of Glancer 338, and from him he passed into the hands of Mr. Frame, Broomfield, and thereafter was owned, according to the particulars in the entry of Superior 836. The sire of Glancer 338 was one of the colts bred by Mr. Archibald Bulloch, and got by Broomfield Champion 95. He is registered as Young Champion 937, and was foaled about 1843, when his breeder was tenant of the farm of Brainzet, in Baldernock parish. Students of pedigree will have observed from what has been now advanced that Drumore Farmer 284 was a somewhat in-bred horse, and older breeders affirm that, in spite of his many undoubted good qualities, he was not free from a disease which authorities are agreed is accentuated by injudicious in-breeding.

Another celebrated horse got by Broomfield Champion was Bowman's Colt 1078, who was foaled in 1841, and was winner of second prize at the Highland and Agricultural Society's Show at Glasgow in 1844. He will be best remembered through his connection with a very fine race of mares in the Croy-Cunningham stud, and the family of Clydesdales represented by the Campsies and the Wellingtons. The Croy-Cunningham Jess was bred by Mr. Alexander Galbraith, and was a well-known figure in the show-ring. Her descendants are numerous and her son, Johnnie Cope 416, the Highland and Agricultural

Society's first prize three-year-old in 1857, was the sire of the first Campsie 119. The blood of Bowman's Colt was also introduced into the Haughhead stud. A colt got by him, named Thompson's Horse of Boghead 1330, was sire of Surprise 846, an animal of great reputation, and the sire of Wellington 906. Bowman's Colt, it is understood, was "foundered" at an early age and passed from the scene with a comparatively limited, but an enduring record of success.

Clyde *alias* Glancer 153 is, however, when the best is said for all else that Broomfield Champion bred, the flower of the flock and the pearl of the tribe. He was probably not a beautiful horse; hence the foregoing similes may be somewhat inappropriate. His praises are not sung in show reports, and indeed those who remember him, while the key in which their remarks are pitched is not a minor one, do not become unduly enthusiastic in his praise. He was a "mickle strong horse." That seems to be the summit of his praise, and from this it is fair to conclude that he at least was virile. If a stallion is not that, he had better never have been born. A masculine female and a feminine male are solecisms in whatever section of animal life they may appear, and the old ruptured horse can afford to lack the meretricious adornment which the show-ring imparts, seeing that he possesses the enduring renown of a tribal head, and that one of the greatest the Clydesdale breed can boast. He was bred by Mr. Forrest, the Hole, Lanark, and there is every reason to believe was the produce of a first-class Clydesdale mare. It is of importance to notice, as corroborating the view just taken of the means that operated to the improvement of the breed of horses in Lanarkshire, that as early as 1823, writers in standard works on agriculture mention the Clydesdale breed under that name as the best-known breed of cart or heavy draught horses. It is, therefore, no straining of the imagination to conclude that a horse so widely recognised as a leading sire as was Fulton's "Ruptured Horse," was out of one of the well-defined tribes of mares generally known as Clydesdales.

From 1844-1850 it may safely be said that the leading honours of the show-yard went to the produce of Clyde *alias* Glancer 153. Several of his sons were even more successful than himself in getting prize stock, and it is an evidence of what has been regarded as his leading characteristic that a greater number of his sons are of historical reputation than there are of his daughters. His name occurs eight times in the 1,044 pedigrees recorded in the retrospective volume of the Clydesdale Stud Book, and only one of these has reference to a mare. Seven stallions recorded were his sons. Baasay 21, so-called because of his broad white face, was a Renfrewshire horse, and maintains his hold on the present-day Clydesdale through the descendants of Barr's well-coupled and typical Clydesdale, the prize horse, General Williams 326, and the descendants of Clark's Prince Alfred 619, a grand big horse that gained first prize at Glasgow, when three years old, in 1870. He was bred by Mr. Allan, Inches, Eaglesham, who owned a very fine tribe of old-fashioned Clydesdale mares. Their influence has been wholly for good in the Clydesdale breed, and if not the most fashionable, they are certainly not the least useful of Clydesdale families.

The other horses got by Clyde *alias* Glancer 153 were much better known than Baasay, and almost all of them were noted prize winners. They were (1) Clyde *alias* Prince of Wales 155, first prize aged stallion at the Highland and Agricultural Society's show at Glasgow, in 1844, and starting with which as a text, a fairly complete history of the modern Clydesdale might be written ; (2) Farmer *alias* Sproulston 290 which, although not so well-known in the show-yards, did splendid work in improving the breed of horses in Bute, and also to some extent in Wigtownshire. (3) Erskine's Farmer's Fancy 298, a prominent prize winner; which served for many years in Kintyre district. He was not quite as good as he ought to have been in the essential ground points, and while his influence was not by any means all for evil, he imparted an inheritance to the Kintyre Clydesdales which they could very well have dispensed with. (4) Muircock 550—a black horse bred by a man celebrated in Clydesdale lore, Mr. John Stevenson, Rakerfield, Beith—travelled in Renfrewshire and Ayrshire, and is generally regarded as having been in point of individual merit the equal of the best of the progeny of the "ruptured horse." He is represented to-day by the sons of the Merryton mare Kate, Luck's All (510), &c., and in the blood of McKean's Prince Charlie 629, and all the other numerous descendants of the Milmain Jess *alias* Beauty 355. (5) Prince Charlie 625, a grey horse that did good service in Wigtownshire, was understood to have come from out of a stud of the old Lanarkshire grey sort : and the last recorded son of the "Ruptured Horse." (6) Barr's Prince Royal 647, was a phenomenal horse in many respects, but one concerning whose merits as a breeding horse there is some diversity of opinion. He was third at the Highland Society's Show at Aberdeen in 1847, and second at the same show at Edinburgh in 1848. In outline he was strong, big and weighty—in fact inclined to be coarse. His merits were great, but he was unequal both in respect of individual points and as a sire. At the same time his progeny were much more distinguished in the show-ring than were the progeny of any other contemporary sire, and indeed all the principal prize-winning mares for some years about 1850 were got by Prince Royal. It is possible that he was not too purely bred, as there was great diversity amongst his progeny in respect of colour. It is even alleged that the only colour which did not appear amongst them was black. He certainly bred several of a chestnut colour, and this goes to confirm the suspicion current in some quarters that there was a flaw, from a Clydesdale point of view, in his dam's breeding. One of the most famous of his progeny was the mare bred by Mr. Kinloch, Kilmalcolm, and owned by Mr. Andrew Logan, Crossflats, Kilbarchan, which gained second prize at the Highland and Agricultural Society's Show at Perth in 1852, and first in the brood mare class at the same show at Berwick-on-Tweed in 1854. Her daughter, got by Clyde 155, was also first in the two-year-old class at Perth in 1852. A full sister of Logan's mare was dam of the horse Sir William Wallace 804, which in a very marked degree was instrumental in later days in improving the breed of horses in Islay. Apart from this connection, however, and the lines of descent that pass through General Williams 326, and Logan's prize mare of Netherton in the Blackhill stud it must be acknowledged that there is but little impression remaining on the Clydesdale breed from Prince Royal. At the same time it is to be observed, that wherever that influence is present there is to be found as a rule exceptional size and weight of bone.

With these foundation strains arising from Clyde *alias* Glancer 153, it may be said that the whole modern Clydesdale race, in so far as it is related to the family of Thompson's black horse, Glancer 335 is identified. But in order to the creation of certain of the best known modern tribes this great trunk line was fused with other outlying strains, and it is needful that something should be said regarding these, and how the fusion was effected.

GALLOWAY CLYDESDALES.

The part of Scotland in which the Clydesdale may be said to have found a second home was the province of Galloway, and especially the county of Wigtown and the Stewartry of Kirkcudbright. It is significant of the rapid strides which that locality has made as a Clydesdale centre, that while in 1845, when the Highland and Agricultural Society's show was held at Dumfries, there was scarcely a single exhibit from the two counties named, the show held in the same place in 1886 would have been like "Hamlet" with the principal part omitted, had the Galloway contingent been absent. It is comparatively easy to fix the date when the modern era in Galloway Clydesdale breeding began, and to name the men who played the leading part in introducing that era; but there were Clydesdales in the Stewartry before the Muirs went, about 1840, from Sornfallo on the slopes of Tinto to the Banks Farm, Kirkcudbright (which has gained a world-wide celebrity in these later days in the hands of Mr. William Montgomery), and there were Clydesdales in Wigtownshire before Mr. Robert Anderson, Drumore, introduced the black mare, Old Tibbie and her neighbour, and the stallion, Old Farmer 576, in 1835, from Lanarkshire. In 1830 Farmer 292, the Balscalloch horse, gained a £30 premium at the Dumfries Show of the Highland and Agricultural Society, and his sire was a Wigtownshire horse, named Clydeside, foaled very early in this century. This horse's name is suggestive of a Lanarkshire origin. It is indeed difficult to account for such a name, except on the supposition that the horse was either bred in Clydesdale or was of the type which had at that early period become identified with the Clydesdale district. There is some reason to believe that Comely, the grand-dam of Garscadden Lovely 40, or her dam was also a prizewinner at the 1830 show, where she is believed to have been purchased by Colonel McDowall, of Logan, an enthusiastic horse-breeder. In this connection we hazard the theory, based on the wellfounded report of the keen rivalry that prevailed between Mr. Anderson, the tenant of Drumore, and the laird of Logan, that the reason for the 1835 excursion of Mr. Anderson into Lanarkshire and Renfrewshire in company with Mr. William Fulton, Sproulston, was to purchase animals in Clydesdale or its neighbourhood that would defeat the Galloway-bred Clydesdales exhibited by Colonel McDowall. In view of the acknowledged success of Mr. Anderson, the probabilities are that the horses reared in Wigtownshire, being what Americans would call "graded up" by means of Lanarkshire stallions from the original Galloway nag of which Shakespeare speaks, were not equal in size and weight to the horses bred in Clydesdale; and by introducing both stallions and mares from Lanarkshire, Mr. Anderson practically introduced a new breed into Wigtownshire. This breed or tribe largely dominated the Galloway draught horses for many years, and formed the foundation on which the splendid modern reputation of the Wigtownshire Clydesdales has been reared.

It is not, however, to be supposed that to this revolution, from which Wigtownshire derived great benefit, it contributed nothing. As has been proved to demonstration within recent years, the most divergent results appear in foals which, starting life on equal terms, have been reared respectively in Galloway and in the West of Scotland. The climate and soil of Galloway are formidable elements in the competition for showyard honours amongst young Clydesdales; and given the possession of the advantages of a Galloway up-bringing, the young Clydesdale starts on its life journey with a considerable advantage over its neighbours.

The early draught horses of Kirkcudbright were, we think, of a lighter type than those of Wigtownshire, and possibly the "grading-up" process, by means of Lanarkshire horses was there begun at a later date than in Wigtownshire. In Sir John Sinclair's "Account of the Husbandry of Scotland," published in 1812, vol i., p. 120, several of his correspondents discuss at considerable length, the relative merits of horses and oxen for agricultural work, and they indicate that the question was still in doubt as to which was the more profitable kind of labour. A pair of horses, they say, cost £64 15s., and three oxen, which were apparently regarded as equal to two horses, cost £42. Horses, they alleged, yearly diminished in value, whereas the oxen increased annually until they were six years old. The latter they regarded as fit for every kind of farm labour, but unfit for walking on turnpike roads; and while the horse was admitted to be superior for harrowing, the ox excelled in the plough. A Penrith gentleman, whose opinion Sir John quotes at p. 126 of the same volume, eulogises the Scottish farm horse, and characterises him as preferable to any he had ever seen in England. They were of greater weight than blood horses, and better adapted for draught. Dr. Singer, who wrote on the agriculture of Dumfries, about the close of last century, states that the work horses in use there were the results of many crossings of different breeds, being larger than the real Galloway, but less than the pad-formed dray-horse of Glasgow and its neighbourhood. He indicates, however, that there was a growing feeling amongst farmers in some districts, for the stronger, if slower, cart-horse; while at the same time, in other places, on account of the necessity that existed for the horse being able nimbly to make his way along narrow bridle-paths, and, mayhap, to lend speed to the smuggler, a dash of the blood of the saddle-horse to give clean limbs and mettle was preferred. He further indicates that the nature of the soil to be wrought had an important bearing on giving direction to the tastes of the farmers in regard to draught-horses. The farmer whose land was light found a comparatively light horse preferable; because he was at once more easily maintained, and performed his work with greater ease and rapidity; while the farmer who had a heavy and deep clayey soil to manipulate called for a heavy, powerful horse to do his work.

A notable book on the early agriculture of Galloway, is the Rev. Sam. Smith's "Survey," published in 1810. The reverend chronicler is eloquent in his praises of the old Galloway nag, and seems almost to regret that the old days of what he calls "predatory excursions" had come to an end. At page 290, he shows himself to have been a Darwinian before Darwin, by his accounting for the hardiness of the ancient Galloway on the principle of natural selection and the survival of the fittest; and at page 269 he discourses thus on the Galloway horses of his own time:—

"It is much to be regretted that this ancient breed is now almost lost. This has been occasioned chiefly by the desire of farmers to breed horses of greater weight and better adapted for draught; and very little value attached, in times of tranquillity, to horses well calculated for predatory excursions. The horses, which in the lower districts are employed chiefly for draught, do not appear to be a distinct breed from the ponies of the moors; but are a variety occasioned by breeding from those of the largest size, and gradually improved, from being kept on the superior pasture. . . . . The breed has seldom been preserved pure, but yet it is not difficult for connoisseurs to distinguish those which have much Galloway blood. They are deservedly held in estimation as being peculiarly adapted for the different purposes of husbandry. They are round in the body, short in the back, broad and deep in the chest, broad over the loins, ... level along the back to the shoulder, not long in the legs, nor very fine in the head and neck: their whole appearance indicates vigour and durability, and their eye commonly a sufficient degree of spirit.

"Though inferior in size to the dray-horses of many other districts, they are capable of performing as much labour and enduring still more fatigue; they are more easily kept and less liable to disease."

Mr. Smith goes on to say that the size of these native horses had been increased by the introduction of well-boned stallions from England and Ayrshire, and, to a less extent, from Ireland.

The raw, strong, and coarse product of this union formed the material on which the influence of the Clydesdale from Lanarkshire was first impressed. Their size was from fourteen to sixteen hands high, and at four years old they were sold at prices from £15 to £50.

The first of the Lanarkshire breed hired by farmers in the Stewartry was Samson 1288, himself foaled in 1827 or 1828, and his grandsire, Smiler, foaled very early in the present century. He was the property of Mr. John Muir, Sornfallo, Lanark, and was bred by Mr. John Paterson, Grange, Pettinain, Lanarkshire, out of a black mare which gained many prizes at Lanark. He was hired in 1831 or 1832, and again in 1833 or 1834. In the intervening year Mr. Frame's Clyde, from Broomfield, was hired, but in view of the multiplicity of horses of that name in early Clydesdale annals, it seems hopeless to attempt to identify this horse. It may, however, be a fair guess that he was Young Clyde 949. Samson 1288, as may be seen from the Stud Book, left a number of colts which were good enough to be kept as stallions in Galloway. One of these was Young Samson *alias* Borgue 1372, a horse bred in the parish whose name he bore. He died when young, but left one foal, Lofty 1187, foaled in 1836 or 1837, bred by Mr. Jas. Muir, Maidland, Wigtown, who sold him to his brother, Mr. John Muir, Sornfallo, by whom he was brought back into Galloway when Mr. John Muir entered on the lease of the Banks farm. Lofty's dam was a mare, Darling, purchased from Mr. Jas. Frame, Broomfield. This Lofty 1187 was sire of another Darling, the dam of another Lofty 456, and he finally was the sire of Jean, the dam of the famous stallion Lochfergus Champion 449, in whose veins the blood of the Wigtownshire and the Kirkcudbright Clydesdales were thus blended, his sire being the Drumore-bred horse Salmon's Champion 737.

THE AYRSHIRE RACE.

We have thus brought down the record of two great trunk lines of Clydesdale breeding to a point which renders it comparatively easy to connect them and show in a large measure the development of the leading families of modern Clydesdales. But there are still more than one family of somewhat ancient lineage whose association with the most brilliant achievements in Clydesdale history has been very marked and to which some attention must be paid. One of these cannot be traced any further back than about 1840, and indeed even that date is somewhat doubtful. The known origin of the tribe was a mare Bell, owned by the late Mr. John Scott, Barr Farm, Largs, Ayrshire. She was mated in 1842 either with the Lanarkshire-bred horse Scotsman 754, or with a two-year-old colt, and the produce in 1849 was Old Clyde 574, the first horse owned by Mr. David Riddell. If Scotsman were the sire of this horse, then the connection of the Ayrshire tribe with the old Clydesdale breed is well established, but if the two-year-old colt were his sire, the origin of the tribe is hardly even within the region of conjecture. Although we are conscious of the strength of the criticism that may be urged against our view, frequent and somewhat extended intercourse with those who are qualified to speak on such matters warrants us in concluding that Scotsman really was the sire of Old Clyde 574, and therefore the first known ancestor on the sire's side of Prince of Wales 673.

KINTYRE CLYDESDALES.

The evidence connecting the native Clydesdale breed of Kintyre with the old Lanarkshire breed is well brought out in the Introductory Essay to the Clydesdale Stud Book.* The later development of the breed in the Peninsula was mainly associated with three horses. To one of these, Farmer's Fancy 298, reference has already been made. The second was a notable old horse Rob Roy 714, whose pedigree is a little like that of one of the characters in "Guy Mannering." He came somewhat irregularly into the world, but on the whole performed excellent service after he had once entered it. It is impossible now to fix definitely the date when Rob Roy was foaled. He was the sire of the dam of Alma 9, and that horse was foaled in 1854. Consequently he must have been foaled as early as 1845 or thereabouts; certainly not later than 1847. He was undoubtedly descended from the old Lanarkshire breed, both his sire and his dam and grand-dam having been purchased in Clydesdale, and his descendants are in a marked degree true to characteristics which have invariably been associated with the Clydesdale name. He

---

* The retrospective volume of the Clydesdale Stud Book was presented to the Clydesdale Horse Society by the Earl of Dunmore, who made special acknowledgment of the labours of Mr. Thomas Dykes, the first secretary, stating that he had collected the traditionary matter embodied in the Introductory Essay.—EDITOR.

formed a link of connection between the Clydesdales of Galloway and the Clydesdales of Kintyre, and there seems to be good reason for believing, on the basis of a comparison of dates, &c., that several of the progeny with which he has been credited in Kintyre should be rather ascribed to his son Rob Roy 2379. His influence on the breed of the present day has been marked and some of the most valued families for breeding purposes are of his tribe. His most notable sons were Merry Tom 536, in his turn the sire of Drumflower Farmer 286, and Hercules 378, the sire of Lord Lyon 489. His daughters were eagerly sought after as highly successful breeding mares. A defect common to the old horse, which has followed most of his tribe with remarkable fidelity, was undue length and easiness or hollowness of the back. Those who are acquainted with the produce of Drumflower Farmer know that his female progeny have a tendency in this direction. Rob Roy, in fact, was locally known in Kintyre as "Sandy Campbell's laigh backit horse," just as the third Kintyre horse, to which we mean to refer Largs Jock 444, was known as "Sandy Campbell's straight-legged horse." The quality of the feet and legs of Rob Roy was thoroughly up to the standard pithily described by his breeder, Fulton, of Sproulston, in the phrase "razor legged," and the same remark holds good in respect of the produce and family of Drumflower Farmer and Lord Lyon. The breadth and flatness of the bones in the majority of the produce of these horses were quite marked, and the excellence of the Lord Lyon hind leg is proverbial. Largs Jock 444 was a horse of the same tribe as the famous Sir Walter Scott 797, the Clydesdale champion at the Highland and Agricultural Society's Show at Dumfries in 1860 and at the Royal International Show at Battersea in 1862, and of course was in this way also of the tribe which afterwards produced Prince of Wales 673. The produce of Largs Jock were characterised by great substance, and one of them, Prince of Kilbride 660, was first three years in succession at the Highland and Agricultural Society's Shows. His daughters also about the same period, that is, late in the sixties and early in the seventies, were quite prominent in the show-ring, and on the whole it may be safely affirmed that few classes of mares were more popular than those got by Largs Jock. On this account previous to the advent of the Stud Book, he divided the honour with Lochfergus Champion of being the innocent cause of a good deal of misrepresentation on the part of horse dealers, a Largs Jock mare or a Lochfergus Champion mare being a conveniently hazy way of setting forth the pedigree of an eligible looking brood mare whose ancestry was unknown.

Horses in Cumberland and Aberdeenshire.

There are two districts which enjoy a very favourable reputation for the character of the draught horses bred in them, which, at the same time, have not from amongst the strains that may be regarded as native to the localities, produced anything exceptional amongst breeding animals. We refer to Cumberland, and Aberdeenshire or the north-east of Scotland generally. The quality of the work-horses reared in these localities is regarded as first-class, and it may safely be affirmed that Cumberland and Aberdeenshire geldings sell for the highest prices of any horses that are in the market for purely draught purposes. Both tribes have clearly-marked lines of descent from the Lanarkshire breed of horses; but with both there has been intermingled the blood of other breeds in a way which has not been equalled in any of the other tribes of Clydesdales to which reference has been made. The consequence in our opinion is the pre-eminence they have gained as work-horses, and their comparative lack of distinction as breeding animals.

The writer of the Introductory Essay to the Clydesdale Stud Book clearly enough establishes the claim of the Cumberland horses to be ranked as a branch of the great Clydesdale family, and subsequent researches have only tended to strengthen this claim. At the same time he indicates that several of the leading horses were believed to have had southern blood in their veins through their dams. It seems to us that it is impossible to think that horses in a district like Cumberland or Westmoreland could, in the absence of stud books and in days before there were railways, have been kept as pure as those in localities like Galloway and the west of Scotland. And the vim and energy of the Cumberland work-horse, in our view, clearly points to an intermingling of hotter blood than that of the comparatively phlegmatic Shire horse with the Clydesdale. This blend is strikingly illustrated in the case of the grey horse, Blyth 79, which, although bred in West Lothian, was largely used in Cumberland. His sire was the Broomfield horse, Clydesdale Jock 172—a half brother of Broomfield Champion—and his dam was Jess, got by Julius Cæsar, a coaching stallion, which served mares in Cumberland and in Scotland. Blyth was one of the best horses of his time, and gained second prize at the Highland and Agricultural Society's Show in 1840, and first in 1841. Curiously his son, Young Blythe, *alias* Sampson 923, also a grey horse, gained first prize at the Highland and Agricultural Society's Show at Edinburgh in 1859, when seven years old, and was for many years located in Aberdeenshire. He was an out and out Cumberland horse. His dam was got by Old Bay Wallace 572, bred in Ayrshire and foaled as early as 1827. He gained premiums, and travelled in Cumberland in 1832, 1833 and 1834, and died in 1838. Both his sire and dam were prize winners. The former, named Clyde, was first at the Highland and Agricultural Society's Show in 1828, the premium horse for Stirling in 1829, and twice winner of premiums in Aberdeenshire after that date. Old Bay Wallace was a whole-coloured bay horse, standing 16-2. The grand-dam of Young Blythe 923, was got by the celebrated Cumberland horse, Old Stitcher 577, whose breeder was Major Millar, of Dalswinton, Dumfries. He was foaled previous to 1815, and was owned successively by Mr. Muir, Sornfallo, Lanark, and Sir James Graham, Bart., Netherby.

He went to Netherby in the year of Waterloo, and had gained a £40 premium at Linlithgow before that; so that it is safe to conclude that he was foaled not later than 1810.

The greatest of the earlier Cumberland horses was, undoubtedly, Old Stitcher's grandson, Young Clyde 949. He was bred near to Hyndford Bridge in Lanarkshire in 1826, and was got by the Cumberland-bred Lofty 453. After passing through the hands of Mr. Frame, Broomfield, he became the property of Mr. Pringle, Jerriestown, Carlisle, and was thenceforth until he died the leading Clydesdale stallion in the North of England. He was a horse of great size and strength, but a little short in rib. He was a dark bay, with a white spot on his forehead, and white hind feet. His fore-feet were dark coloured. There were few better horses.

As a show horse, the best known of the older Cumberland Clydesdales was, undoubtedly, Phillips' Merry Tom 532. This was a grey horse, foaled in 1848; winner of first prize at the Glasgow Show in 1852; first at the Highland and Agricultural Show at Berwick-on-Tweed in 1854; and first at the Royal Agricultural Society of England at Carlisle in 1855. He was the premium horse for the Glasgow district in 1852, and has been described as the best looking and the worst breeding stallion that ever gained the Glasgow prize. In all probability there is something of prejudice in the second item of this description, but we have never met with anyone who had aught but praise for his individual merits. He left few foals, but amongst them was a second prize winner at the Highland and Agricultural Society's Show. It is interesting to notice that Merry Tom was descended on both sides from Old Stitcher. His sire was Merry Farmer 531 by Young Clyde 949, and his dam was Jean by Batchelor 1056. Batchelor was a brown horse foaled about 1829, and his sire was Scotch Miracle 750, foaled in 1820, and got by Old Stitcher. Merry Tom was a singularly rich looking horse; and there is some reason to believe that Maggie *alias* Darling, first prize brood mare at the Highland and Agricultural Society's Show at Glasgow in 1857, and dam of General 332, was got by him.

The Cumberland line of Clydesdales was represented for many years by one of the families in the celebrated Keir Stud. They have, however, long been absent from that collection; and the best known representative of the tribe in recent years was the Kintyre horse, Lorne 499; which left stock with first-class tops and good legs, but, unfortunately, very deficient in their feet and pasterns. As if these faults were not bad enough when found in one, he was one of the most prolific sires Scotland has ever produced. When in the Keir Stud, the merits of the representatives of this particular tribe were briefly summed up in the words: "Individually good, big, useful horses, but most disappointing as breeders."

Another link of connection between the horses of Cumberland and the north-east of Scotland is to be found in the fine race of the grey Glenelgs. The best stallion of this tribe was Glenelg 357, winner of first prize at the Highland and Agricultural Society's Show at Dumfries in 1845, beating the Kintyre Farmer's Fancy 298. He travelled latterly in the north of Scotland, where his descendants can still be traced. He was a grandson of Young Clyde 949, and like Merry Tom was descended on the dam's side from Old Stitcher through Scotch Miracle. He was owned in 1858, when 19 years of age, by E. and M. Reed, Beamish Burn, Durham.

Our view, that the Cumberland and north of England horses were somewhat of a mixed breed, has in the main been a matter of conjecture based on the geographical position of the breeding area. We are, however, able to point now to a notable well-bred Lincolnshire horse as one that was used in the county for some years. This was Farmer's Glory, owned by John Robinson, Wallace Field. The card of his son Royal Farmer's Glory 5312, contains the following eloquent tribute to his worth:—"Farmer's Glory was a never-to-be-forgotten Lincolnshire bred horse, the property of John Robinson, Wallace Field. He gained first prize at the Royal Agricultural Society's Exhibition at Windsor, and in the following year the Manchester and Liverpool prize; and at Ayr in 1857, a prize of £50, and in

7

1858, a prize of £50. His sire, Seward's horse, Major [1447 (S.S.B.),] gained first prize two years at Wisbech; grandsire, Bingham's England's Glory [705], took the prize three years at Lincoln; great grandsire, celebrated horse, Wiseman's Old Honest Tom [1060], purchased by Mr. Wood, Cottenham, for 400 guineas."

West of Scotland readers will have ere this identified the foregoing reference with Andrew Hendrie's Farmer's Glory, which travelled two seasons in Ayrshire, in the first doing splendid service, and in the second, with a result that was practically *nil*.

The view that the older race of Aberdeenshire draught horses were a somewhat mixed breed is not so much a theory as the result of an examination of the history of horse-breeding in Aberdeenshire during the first half of the present century. The characteristics of the Aberdeenshire geldings are their endurance, cleanness of limb, hardness of bone, and size combined with activity.

Clydesdale stallions tracing directly from the Lanarkshire breed were used at an early period in this century in Aberdeenshire and the neighbouring counties of Kincardine and Banff. In 1823, a horse named Young Glancer, owned by Mr. Thompson, Glasgow, travelled in the counties. He stood barely 16 hands high and was first at Hamilton as a three-year-old. His sire was Glancer, and there is good reason to believe that this was Thompson's Black Horse. From 1846 to 1854, Young Champion of Clyde, a horse foaled in 1840 and a prize winner in the West of Scotland, travelled as the property of Mr. Milne, Mill of Ardlethen, Udny. A dark brown horse named Farmer's Fancy, foaled in 1847, and got by Old Glancer, travelled for some years after 1851. He was owned by Mr. J. Ironside, Bruxiehill. The old horse Justice 420 was, under the name of Emperor, travelled for some years by Mr. Peter McRobbie. A strong but coarse line was introduced by an Earl of Kintore in Old Samson, the sire of Noble 1230. He was a black horse, standing over 16 hands, with a stout, well-coupled body, but curly-haired legs. The well-known Mr. Barclay, of Ury, introduced a dun-coloured chesnut—a strong-boned big horse bred in Dumfriesshire—about 1836. Two of his sons, Black Tom and Rattler, bred by Mr. Henderson, Savoch, Foveran, were for many years popular horses in the locality. The reputation of these animals, however, is completely eclipsed by the two horses Grey Comet 192 and Lord Haddo 486. These were without doubt the most distinguished, and left the best stock of any Clydesdale horse at that time in use in Aberdeenshire. Comet was foaled 1849, and was first at the Highland and Agricultural Society's Show at Inverness in 1856. He was a horse of excellent merit, very thick and short-legged, and his reputation is still green. He was destroyed at Dalkeith in 1866, having got his leg broken by a kick from a mare. His descent from the Lanarkshire breed is very clear on his sire's side, and the name of his dam, Lanark, suggests that she too came from Clydesdale.

Lord Haddo 486 was a younger horse than Comet. He was foaled in 1853, and gained second prize at the Highland and Agricultural Society's Show at Aberdeen in 1858, and died in 1874 or 1875. He was got by the Ayrshire Old Clyde 574, and judging from his portrait, which formed the figure-head of stallion cards in Aberdeenshire for many a day, he was of prime quality, and built as a draught horse should be built.

While the flow of English stallions to Scotland swept past the Galloway and Clydesdale preserves of the native breed, it made quite an invasion of Aberdeenshire. And not only were what are now known as Shires imported north, but about 1816 we find that Mr. Ferguson, of Pitfour, kept Suffolk stallions. They are said to have been very different from the modern Suffolk, having hair to the knees in front and to the point of the hocks behind, and it is therefore possible that they were not real Suffolks at all. At any rate Mr. Ferguson had a high opinion of them, and was so anxious that his tenantry should patronise them that he fixed the terms to them at 2s. 6d. each, and 13s. each mare to all others not his tenants. In 1818 and for some years afterwards, a Suffolk horse named Captain was kept at Newe, and about the same time a strong dark brown horse with black legs, of the Durham breed, was kept at Mains of Gight, Fyvie. Suffolk horses continued to be used the district for some years, the last of them being a chesnut horse named Suffolk Champion, which ended his days somewhat ignominiously driving coals.

A writer in the *Aberdeen Journal* of 1st March, 1880, to whom we are indebted for some of the foregoing information, expresses the opinion that the extinction of the race of Suffolks was of little consequence, as they neither wore well, nor suited the climate of Aberdeenshire. It may be so, but we should not be surprised, if the endurance and cleanness of limb so marked in Aberdeenshire draught horses were proved to be the result of the use of these or some of these East Anglian stallions.

Several good Shire horses were in use in Aberdeenshire in the first half of this century. One of the best of these was Stanmore, bred in Lincolnshire about 1832. His importer was Mr. Boswell, of Kingcausie, and he was greatly admired during his lifetime. Obviously the breeders of those days were not very particular about colour, for Stanmore was quite white. He is described in these terms:—" He was a very complete horse, perfectly white, with a wide forehead, large open nostrils, and small sharp ears. He had a curly mane reaching past his knees, and his tail almost touched the ground. He stood only about 16 hands, but was very strongly made, his worst fault being that he was rather straight in the hind legs." The influence of the blood of Stanmore is chiefly to be traced through Rory O'More 718, a grand grey horse, 16·1, and of the best quality. Although he left many foals, none of his produce was regarded as quite equal to himself.

The only other English horse to which we will refer is Black's Champion of Cairnleith, a grey horse foaled in 1838, and bred by Mr. Jacob St. John Ackers, Prinknash, Painswick,

Gloucestershire. This was a magnificent animal, and by general consent one of the most valuable stallions ever used in Abérdeenshire. He bred many foals during his long life, and altogether was a horse that did a power of good. It is however to be noticed that there is very little of his influence to be traced in the blood of leading Clydesdales during the past twenty years, the most notable quarter in which it appears being in the Mearns mare Jean by Eclipse 268, the dam of What Care I 912, and of something like half-a-dozen other stallions.

Modern Clydesdales and their Characteristics.

The sketches that have been given of the various tribes and families of Clydesdales will have rendered it comparatively easy to tell the story of Clydesdale breeding during the past ten years. A chapter might have been devoted to the work of Mr. Lawrence Drew, and the effort which he made to found a distinct class of Scotch draught horses through an amalgamation of the modern Clydesdale and the modern Shire; but his untimely death in 1884 cut short the experiment at a time when his friends had hoped that he was about to establish, by practical results, the truth of the theory that he had advanced, and consistently adhered to. That theory, briefly expressed, was that the Clydesdale and the Shire are one and the same breed, and that the best draught horse is to be bred by a fusion of the two. The two divisions, he argued, were not representative of two breeds, as the Aberdeen-Angus and the Shorthorn are two breeds, but of two wings or sections of the same breed as the Booth and the Bates cattle are alike Shorthorns. No one who attended Scottish Shows during the years from 1875 until Mr. Drew's death, could deny that he bred some marvellously good horses and mares from Prince of Wales 673, and what are now well known to have been a particularly good class of Shire mares. The important question, however, was, "Will these excellent results be perpetuated by a continuance of the same methods of breeding?" and that is precisely the question which it has been rendered impossible to answer satisfactorily on account of the abrupt termination of Mr. Drew's career. The results of the past six or eight years have gone to show that several of the animals—male and female—bred by Mr. Drew or on his lines, have done very well as breeding stock when mated with reasonably pure-bred Clydesdales—that is, Clydesdales in which the blood of the old Lanarkshire breed was predominant in a marked degree.

Leading Tribes.

Taking the results of the showyard for the six years, 1886-91 inclusive, as a fair means of knowing the principal factors in modern Clydesdale breeding, we find that the great majority, indeed all the successful sires, are easily summarised under six heads: Darnley, Prince of Wales, Lord Erskine, Drumflower Farmer, Old Times and Lord Lyon. Amongst the first dozen sires represented by prize stock at the principal shows in these years, there is not a horse which cannot without violence be easily included as of one or other of these families. The Darnley interest, during the period mentioned, is strongest. He himself heads the list of winning sires in 1886, 1887, 1888 and 1889, and in three of these years his son Macgregor stands second to him, while not less than four and as high as seven of the successful sires in each of the six years are either his sons or grandsons. This speaks strongly in favour of a high uniformity of excellence in his stock, and it is of importance therefore to see of what constituents his own pedigree is composed.

Darnley 222 was bred by the late Sir William Stirling Maxwell, Bart., at his Keir Stud Farm, in 1872; and was owned by Mr. David Riddell from the time he was three years old. He had a very distinguished showyard career up to his twelfth year, when he was champion male Clydesdale at the Centenary Show at Edinburgh. He died on the 30th September, 1886. His sire was Conqueror 199, a Kirkcud-

From a Photograph by C. Reid.]

CLYDESDALE STALLION, FLASHWOOD 3604.
Winner of First Prizes at H. and A. S. Shows.
Owned by Messrs. A. & W. Montgomery.

bright-bred horse, somewhat undersized, and having somewhat defective action behind—in other words he walked wide behind. He was got by the massive, big and somewhat "raw" Clydesdale, Lockfergus Champion, whose blood constituents have already been described. The dam of Conqueror was a Galloway Clydesdale—that is, she gave evidence of having been "graded up" after the manner already briefly described, from the native stock of Galloway. Her sire's name does not appear in the Stud Book, but we have good reason to believe that he was Jack's the Lad 400. She was a well-known mare in her time, and of so much note that there is to be seen to this day in a "bog" in the croft, on the farm of Culcaigrie, in the parish of Twynholm, a moss oak which marks her grave. In every way she was a good example of the older race of Galloway Clydesdales, having good feet and legs, and a very hardy, durable constitution. Conqueror 199, her son, was the Dunblane, Doune, and Callander premium horse in 1871, and as Keir Peggy 187, the dam of Darnley, had been served all season by the Keir stud horse, and had not been stinted, as a last resort, and with no other thought than that of getting a foal out of her somehow, she was, at the close of the season, mated with Conqueror. The result was the greatest of all her produce, and one of the greatest stallions of the century—Darnley 222.

Keir Peggy and her tribe have a long and honourable Clydesdale history. She was a dark bay mare of great size, weight and strength, foaled in 1860, and bred by Mr. Hugh Whyte, Barnbrock, Kilbarchan, Renfrewshire. Her prize record in her youth was a very formidable one, and she was widely and favourably known as the "Barnbrock filly." She was bought for Sir William Stirling Maxwell, Bart., by that enthusiastic Clydesdale fancier, his factor and friend, Mr. Alexander Young, and her career as a brood mare was wholly confined to the Keir Stud. She died at Keir on 24th November, 1888, having produced ten foals. Of these, three, the stallions Pollock 592, Newstead 559, and Darnley 222 were first prize winners at the Highland and Agricultural Society's Shows.

The sire of Keir Peggy was one of the most remarkable stallions that ever was foaled. He was a comparatively insignificant beast himself, and his local *sobriquet* of "Logan's Twin" sufficiently indicates the cause. The other side of his history is equally well brought out by his recognised name Samson 741, for he was a veritable giant in respect of his breeding record amongst Clydesdale sires. His breeder, Mr. Andrew Logan, Crossflats, Kilbarchan, Renfrewshire, deserves to be ranked amongst the very foremost breeders of Clydesdale horses. Between the years 1850-1865 no name more honourably figures in the prize lists, and it says much for his success that he bred Samson. At the Perth Show of 1852, he showed very successfully, but the mare which did him most service was the first prize yearling filly at that Show, then shown by her breeder, Mr. Jack, Balcunnock, Campsie. This filly was got by Hilton Charlie 381, a son of Samuel Clark's Clyde 155, out of a mare by Clydesdale Jock 172, and her dam was a chestnut mare of unknown pedigree, bought at a Falkirk Tryst. She was purchased later by Mr. And. Logan; was second in the three-year-old class at the next Show of the Highland Society, held at Berwick-on-Tweed in 1854, and became dam of Samson *alias* Logan's Twin 741, and a much more noted horse as a prize-winner, Logan's Lord Clyde 477. Samson was for a time located at Mr. Calder's farm of Colgrain, in Cardross, Dumbartonshire. He was stud horse at Keir for several years, and was also at one time owned by Mr. Riddell. He died in the possession of Mr. Oliphant Brown, Shiel, New Galloway. Not a few of the best breeding mares in the Stewartry were got by him: and, taken all in all, he may be described as the most impressive Clydesdale sire up to his own time, and many of the most impressive bred since have been descended from him. Doubtless his dam owed not a little of her success to the strength of the old Hilton blood inherited from her sire. But that the success of Samson is not altogether to be attributed to his dam is clear when the fact is recalled that her other son, Logan's

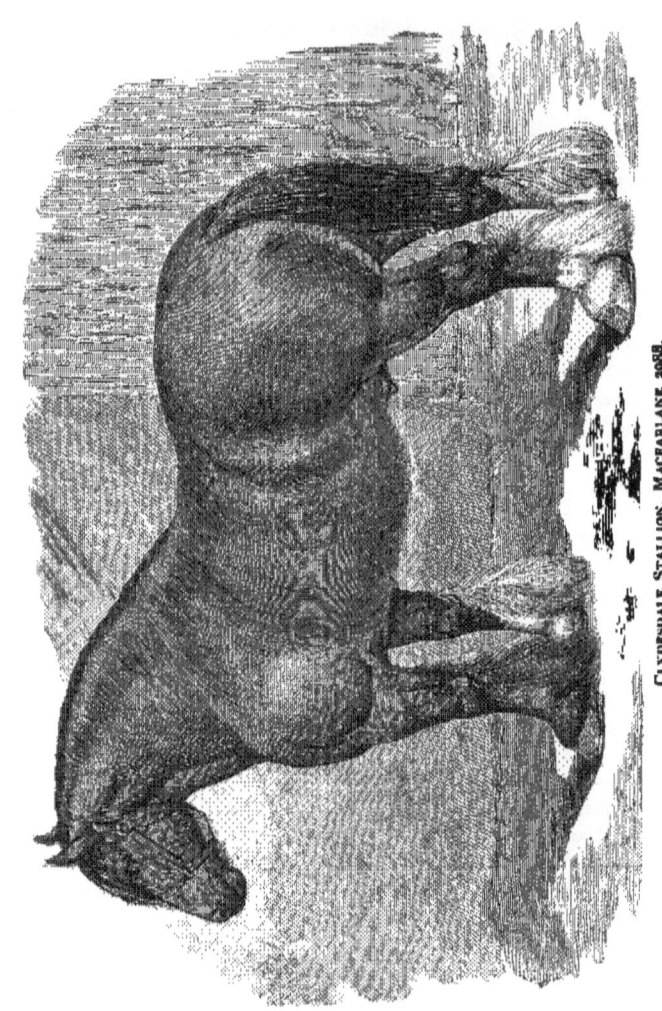

CLYDESDALE STALLION, MACFARLANE 2968.
Winner of First Prize at H. & A. S. Shows, 1885 and 1886.
Owned by Mr. Andrew Montgomery.

Lord Clyde, far surpassing Samson as a show horse, and winning many prizes, was inferior to him as a breeding horse, and consequently some credit must be given to Samson's sire, the Glasgow premium stallion of 1856, Lofty 455. This fine horse was bred in Kintyre, out of an old stock, and his sire was Erskine's Farmer's Fancy 298, to which reference has more than once been made. It will thus be seen that Samson was a distinctly in-bred horse. His grandsire on the top line, Farmer's Fancy, and his great grandsire on the dam's side, Clyde 155, were half-brothers; both, as we have already seen, having been got by Clyde *alias* Glancer, the Ruptured Horse. This fact is specially worthy of notice, because no less than three of the heads of families specified by us as leading amongst modern Clydesdales, are the produce of mares got by Samson. These are Darnley, Prince of Wales, and Old Times.

The dam of Keir Peggy 187 was Jean, bred by Mr. James Holmes, Auchincloich, Kilmalcolm, Renfrewshire. She was got by Erskine's Farmer's Fancy 298, and her dam was reared off a stock of Clydesdales kept on the Sclates Farm, Kilmalcolm, whose history dates from the beginning of the present century at least. It will thus be seen that the pedigree of Darnley dates from an early historical period, and except the origin of the Balcunnock mare is easily traceable to a Clydesdale original. Whether the Balcunnock chesnut mare was, as some think, of English extraction is a question which there are now no data to settle. The colour certainly lends strength to the supposition.

The characteristics of Darnley's family are well-known to all frequenters of Scottish showyards. Generally they are well-coloured, inclined to be dark rather than light brown or bay, and dappled, with few white markings, but with almost invariably at least one white foot and a white mark of some kind or other on the face or forehead. The action of the old horse himself when walking was as near perfection as one could wish for. He took a long, steady step, and got over the ground with marked celerity. His trotting action was defective. He cast or dished his fore-feet, but moved well behind. His feet were fully up to the standard requirements, and his bones were of the best wearing material. He was as clean in the limbs when he died as a two-year-old colt. He had long pasterns, and indeed, as it is expressed in Scotland, he was uncommonly good at the ground. His weak points were in the development of his forearms, which were somewhat lacking in muscle, and his thighs were also open to the same objection. He had a splendid formation of shoulder and neck, well-rounded barrel, and a good straight back, but drooped a little in his quarters. His head was considered to be rather small and pony-like, and his ears especially were thought to be too small. His own characteristics have been reproduced with marked fidelity in his descendants to the third and fourth generations, and have consequently been the means in many respects of improving the breed. He—and this is generally true of his descendants—arrived at maturity slowly; but when fully grown and on his season, he weighed over 20 cwts.

Prince of Wales 673, the great rival head of a tribe, had a much longer life than Darnley. He was foaled in 1866 and died December 31st, 1888. His breeder was Mr. James Nicol Fleming, then of Drumburle, Maybole, Ayrshire. He was got by a Highland Society first prize stallion, and his dam was a Highland Society first prize mare. Not only so, but his sire, General 322, was got by a Highland and Royal Agricultural Society first prize stallion, and his dam was a Highland Society first prize mare. It is a curious coincidence that both of his grand-dams were grey mares. As a show horse Prince of Wales may be said in his prime to have been practically without a rival. No doubt he was placed second on one occasion, just as Darnley was, but although both decisions may have been correct at the time they were given, no one presumes to affirm that either Prince of Wales or Darnley was inferior to the respective horses which beat them. Prince of Wales was owned until he was three years old by

CLYDESDALE STALLION, PRINCE OF CARRUCHAN 8151.
Winner of First and Champion Prizes, Highland and Agricultural Society's Show, 1893, &c.
The Property of Messrs. P. & W. Crawford.

his breeder. He then passed into the hands of Mr. David Riddell, Blackhall, Paisley, who sold him to the late Mr. Drew. He remained in that gentleman's possession until his death, and at the Merryton dispersion sale held consequent on that event, on April 7th, 1884, he was sold by public auction for 900 guineas, and again became the property of Mr. Riddell, in whose possession he died.

The sire of Prince of Wales was General 322. He was a big strong horse, bred by Mr. Thomas Morton, Dalmuir, owned by Mr. Riddell, and exported to Australia when rising four years old. Although, as we have said, a Highland and Agricultural Society's first prize winner, he is not remembered for anything but the fact that he was sire of Prince of Wales.

His sire was the celebrated Sir Walter Scott 797, a son of the old horse Old Clyde 574, referred to in a previous part of this paper. He was the most active, neatest and most stylish horse of his time and possibly of any time. The gaiety of his action is proverbial, and although not a horse of the largest size or greatest weight he was so evenly balanced that none could gainsay his title to first rank. He was placed second once—his successful opponent being Barr's General Williams 326. Sir Walter Scott is one of the most purely-bred Clydesdales the records of horse breeding can boast.

The dam of General 322, was Maggie, *alias* Darling, known locally as the Wellshot Grey Mare, from having been owned first in the West of Scotland by Mr. Buchanan, Wellshot, Cambuslang. She was a mare of great weight and many good qualities, and because of her relation to Prince of Wales, her antecedents have given rise to a good deal of controversy. Various theories have been advanced as to her origin—and the fact that the appearance of Prince of Wales 673, especially about the head, indicated an English strain in his blood has, doubtless, had something to do with the formation of an opinion held in many quarters that both of his grand-dams came from the south. On the whole, having heard all the theories that have been advanced regarding the dam of

General, and seen the evidence by which they are supported, the writer is disposed to attach most importance to the one which traces her origin to Cumberland and names her sire as Merry Tom 532. The authority for this statement was the late Mr. Wilson, farm manager at Wellshot, who purchased the mare from the late William Giffen, horse dealer, Newton Mearns, Renfrewshire, and was aware at the time of purchase that Mr. Giffen had got her in Dumfries. It may be of interest in this connection, as showing the significance of the terms used in the West of Scotland regarding horses, to remark that Merry Tom himself by the older breeders in the Glasgow district, was always called "the English horse that came from Carlisle."

The dam of Prince of Wales 673, was named Darling. She was a magnificent dark-coloured mare, with the best of feet and legs, and lived to a good old age, and died at Merryton. She was bred by Mr. Robert Knox, Foreside, Neilston, and was got by Samson, alias Logan's Twin, being thus half-sister to Keir Peggy. Hawkie, her full sister, was a Highland Society prize winner like herself, and was dam of the well-known Old Times 579, about which we have yet more to say. The dam of Darling and Hawkie was the grey mare Kate, which Mr. Knox purchased from Mr. William Giffen, horse dealer, Newton Mearns, who purchased her in Dumfries. Like the other grand-dam of Prince of Wales, various theories have been advanced regarding her antecedents, but nothing certain is known. If she was, as is believed by some, a mare purchased in the Midlands, she was of the same type and character as some of those which the late Mr. Drew purchased in later years about Derby. Another opinion that is held is that she was bred in Dumfriesshire, and that her sire was Blyth 79. She was a somewhat quick-tempered mare, and consequently was not popular in work, but she bred several first-class foals.*

---

* Mr. Nicol Fleming (the breeder) and Mr. Lawrence Drew (the owner) of Prince of Wales placed on record their belief that both the grand-dams of that horse were Shire mares.—EDITOR.

Prince of Wales was a dark brown horse, with a white stripe on face; the near fore-foot and fetlock and the off hind-foot and fetlock were white, as was also the near hind leg half-way up to the hock. His off fore-foot and leg were wholly dark-coloured. At the ground, in respect of feet and pasterns, no possible fault could be found with him, and so perfect was he that at these parts he has always been regarded as the model. He had broad, clean, flat bones, with the sinews very clearly defined. His hocks, and consequently his hind legs, were too straight, and this was his worst defect. The formation of his fore-feet and legs was perfect. His neck and head were carried with great gaiety and style; his shoulder was set at the proper angle and his back was firm, while his ribs were well sprung from the back, but not deep enough, especially behind. His quarters and thighs were well developed, and, indeed, the general outline along the top was very pleasing. His head, as we have said, was a little "sour," that is, inclined to be Roman nosed: it was of proper length, but not as wide between the eyes as the typical Clydesdale head. The most striking feature of all in Prince of Wales was his marvellous action, and this was all the more remarkable in view of the straightness of his hocks which one would have thought would have been inimical to easy movement. Both at walking and trotting pace the action was perfect. This feature generally characterises his descendants, and the straightness of hock is sometimes also apparent, accompanied in not a few cases by the action called in Scotland "going wide behind." The parental formation of head is also unduly prominent amongst his progeny and their descendants, but it is a feature which in many cases appears less marked as time goes on. The family are, as a rule, characterised by a striking immunity from hereditary disease, and this, combined with their fine wearing qualities and generally easy action, has caused them to be highly popular.

Lord Erskine 144 is a much younger horse than either of

the two that have been named. He is, although remotely, in touch with Prince of Wales, through Sir Walter Scott being a common ancestor, and with Darnley, through Lochfergus Champion being a common ancestor, to all intents and purposes the head of a distinct family of the Clydesdale breed. Lord Erskine was foaled in 1879, and was bred by Mr. W. S. Park, Hatton, Bishopton. He was got by Boydston Boy 111, a short-legged typical old-fashioned Clydesdale stallion, a prize winner at the Highland and Agricultural Society's Show, and bred from a combination of Wigtownshire and Kintyre blood. The dam of Lord Erskine was Hatton Bella 626, a Renfrewshire prize mare, got by the noted horse, Time o' Day, 875, and his grand-dam was a good old Clydesdale sort, got by Young Wattie 1042. The Lord Erskine family are marked by these characteristics : substance, shortness of limbs, breadth of bone, fairly good feet and pasterns, excellent tops, exceptionally good quarters and thighs, and first-rate typical Clydesdale heads. A goodly number of them are somewhat light in colour, and perhaps some are rather straight in the formation of their hocks, and move somewhat widely behind. They are a particularly healthy tribe, and most valuable for crossing purposes, on account of their weight and substance.

Farmer 286, or as he is usually designated, to distinguish him from the numerous other horses of the same name, Drumflower Farmer, was an out-and-out Galloway Clydesdale. He combined the best of the Wigtownshire blood with the advantages of a dam bred and reared in Kirkcudbright, and his general characteristics continue to this day to be common to Clydesdales bred in the south of Scotland. He was a notably in-bred horse, and yet of all the horses that have been specified as heads of families, he was the strongest, biggest, most powerful, and altogether of the greatest substance. This we think is to be attributed to the circumstances of his breeding and rearing. Whatever defects Farmer's stock and descendants may have had, lack of substance was and is exceedingly rare amongst them. He was bred by Mr. Robert Frederick, Drumflower, Dunragit, and was foaled in June, 1869. He passed through various hands, but died in the possession of Sir Robert Loder, of Whittlebury, Bart., in 1883, who purchased him at the Cambus dispersion sale in July, 1881. He was a bay horse, with white markings on face and feet and legs. His sire was the Drumore horse, Merry Tom 536, got by Rob Roy 714, out of Tibbie, the dam of Victor 892, and Salmon's Champion 737, and his dam was Mary, bred by Mr. W. Rain of Miefield, Kempleton, Twynholm, and got by Lochfergus Champion 449, out of a mare owned by Mr. Rain, and bred in the Stewartry. Lochfergus Champion, as has already been pointed out, was got by Salmon's Champion.

Merry Tom was the least-known, but in the opinion of the best judges, the most successful breeding stallion produced by Drumore Tibbie. He inherited his sire's fault of an "easy" or hollow back, but his progeny were wonderfully sound animals, with excellent feet and limbs, and favourite breeding stock. All the three sons of Drumore Tibbie that have been named, travelled in the Stewartry, and there is a consensus of opinion that Merry Tom was the best of the three. The breeding of Farmer is an example of the closeness of mating that was not uncommon in the Stewartry, when, in the absence of a stud book, the premium horse was engaged for service of mares, without much regard to relationship. It is almost certain, in harmony with the views that then prevailed, that had it been known how closely related the three Drumore horses and Lochfergus Champion were, they never would have been patronised as they were, and such horses as Farmer never would have been bred. And yet there seems to be no reasonable doubt that the breed of mares, of which Kirkcudbright has long been able to boast, would never have existed but for this unwitting close-breeding. Drumore Tibbie, the dam of Merry Tom, was a remarkable mare. She was invincible in the showyards, and bred at least one other

stallion besides the three named above—Prince Charlie—
which has not been registered, because he was exported to
Australia when two years old off. He is, however, admitted
to have been the best of the lot, and was first at the Highland
and Agricultural Society's Show at Glasgow in 1857. Her
dam, Drumore Susie, was even a greater wonder. She was
in her dam, Old Tibbie, the black mare which Mr. Anderson
purchased from Mr. Young, Brownmuir, Lochwinnoch, at the
same time as he purchased Old Farmer 576, at the date of
purchase, and her sire was never known. She was foaled in
1836, and died in 1859, from first to last giving birth to nineteen foals, and twice she produced twins. Old Tibbie herself
was first prize winner at the Highland and Agricultural
Society's Show, held at Ayr in 1835. Bred so deeply from
first-class blood, it is not to be wondered at that these
Drumore horses should have been so greatly sought after, and
that their influence should be so enduring.

Farmer 286 was a horse, marked, as has been said, by much
substance and weight. He travelled in several widely separated districts, and in every case his female progeny have been
in high favour. As breeding stock they are almost unrivalled,
and many good—indeed, first-rate—animals have been out of
them. The worst defects to which the tribe are liable have
been inherited from their head. One of these has been men
tioned—that of being rather long and easy in the back. This
is a Rob Roy feature which reappears often when it is not
wanted. Another failing which occasionally appears, is that
of having rather thin and soft feet. The hoof is not formed
of the toughest material, and its formation is possibly sometimes not compact enough. The prevailing colour of the
tribe is a good red bay—what the Americans aptly term
"mahogany bay," and although four white legs and a white
face are not uncommon markings, they are not unduly
prominent.

Lord Lyon 489 was a cross-bred stallion. He was got by
Hercules 378, out of an English, and we presume a Shire,

mare. He was bred, and with the exception of a short interval of three seasons, 1871-2-3, owned from his birth in 1867 until his death in September, 1881, by Mr. John McMaster, Culhorn Mains, Stranraer. He was a masculine, breeding-like roan horse, and when mated with Galloway mares, especially those got by Victor 892, Glenlee 363, and Farmer 286, he bred some of the most extraordinary showyard animals seen during the years from 1874 until 1883. His progeny that acquired distinction were mainly females, and of almost all of them it may be said that they inherited the typical Galloway Clydesdale features of their dams, intensified by the mysterious property which marks the dividing line between the good beast and the prize winner. In one respect Lord Lyon undoubtedly added something to the betterment of the Clydesdale. His progeny were marked by a peculiarly pleasing formation of hind leg, and this has continued amongst their descendants. Amongst prominent prize horses of recent years, bred from a combination of Prince of Wales, Darnley, and Lord Lyon blood, the influence of the Culhorn Mains horse has been very clearly seen in the absence of the defects in thighs and hocks to which we have referred in our notes on the two leading families, and in the presence of that almost ideal formation of hind leg by which Lord Lyon's stock were distinguished. Perhaps the chief defect in the Lord Lyon tribe has been a tendency to softness. At the same time several of his daughters are amongst the healthiest, hardiest and soundest mares in Galloway and in Angus.

The Old Times tribe have contributed as many superior mares to the Clydesdale breed as any one, or perhaps as any two, of the families previously specified except that of Darnley. He was bred by Mr. Robert Knox, Foreside, Neilston, Renfrewshire, and was foaled in 1869. If he is dead the event must have taken place in Wigtownshire about two years ago (1890). He passed through the hands of various owners, and his stock are to be found in many districts of Scotland, but chiefly in Ayrshire, Wigtownshire, and Kintyre. We have already sufficiently indicated his breeding on the dam's side. He was full cousin to Prince of Wales 673. His sire was a noted horse, Lord Clyde 478, winner of first prize at the Highland and Agricultural Society's Show at Aberdeen in 1868, and soon thereafter exported. In his veins there blended the blood of an excellent English horse, Proudfoot's Emperor, bred in Cambridgeshire, and a capital old Clydesdale tribe, owned by Mr. Duncan Macfarlane, Torr, Helensburgh. Old Times gained first prize at Glasgow when three years old, and was generally regarded as the best three-year-old stallion of his season. He was a horse of great substance, with wide, open feet; broad, big, flat bones; good, strong, clean joints; powerful forearms and thighs, a fair shoulder, a good Clydesdale head, a long easy back, a tendency to flatness of rib, and what is called in Scotland, "lowness behind the shoulder." He had only fair action, and moved in front with a peculiar step, putting down the heel in a loose, indefinite kind of way. There have been, and are, one or two stallions amongst his sons that have bred fairly well, but, notwithstanding his undoubtedly strong masculine appearance, there never was a really first-class horse. With his female progeny the case is entirely different. The phenomenal breeding mare of the race, Duchess of Challoch 4780, was got by Old Times, and although all of his produce are not like her, they are as a class remarkably safe breeding mares, and some of them have taken the highest showyard honours. One defect in his male progeny which was not at all marked amongst his females is a tendency to softness both in feet and in general health. The character of his stock and the stock got by Drumflower Farmer suggests that what would be regarded as serious defects in a stallion are comparatively trivial and even in a modified form advantageous in a mare. This is true, for example, of length of body. One great attraction these two tribes of females have is that they are long below and have plenty of room for carrying healthy, strong foals. Hence it is that Farmer and Old Times mares are always in demand.

Clydesdale Stallion, Top Knot 6360.
Winner of First Prize, Glasgow Stallion Show, 1869
The Property of Mr. A. B. Matthews.

MISCELLANEOUS TRIBES.

It may, perhaps, be thought that in restricting the title "Leading Tribes" to the families above reviewed an undue importance is attached to them. But the most striking fact in connection with these tribes is, that even had the basis of our classification been the whole of the sires of winning Clydesdales at the principal shows during the six years stated, the supremacy of their families would not have been seriously impaired. At the same time, there are one or two other lines of breeding which have rendered good service to the Clydesdale cause which merit a concluding passing reference. Chief amongst these are the Topsman 886 family, and the Lochburnie Crown Prince 207 family.

Topsman 886 was an Aberdeenshire-bred stallion, got by a son of the Old Clyde 574 that stands amongst the first of the Prince of Wales tribe, and having as his dam a mare with a curious history. Her name was Jane, and she was foaled in the possession of Mr. William Wilson, Whiteside, Alford, on which farm she remained until she died. She gained numerous prizes, including second at the Highland and Agricultural Society's Show at Aberdeen in 1858. Her dam was a grey mare Peg, purchased in a Glasgow market when in foal with Jane, and sold as being stinted to a horse named Samson. On the one hand, it has been asserted that this was a Clydesdale horse—it was so represented at the time of purchase; and on the other, it is alleged that he was the well-known English horse Bryan's Sampson, which is said to have gained first prize at the Royal Agricultural Society's Show at Shrewsbury in 1845. Whatever the sire of his dam may have been, there is no doubt that there were in Topsman features that are more commonly associated with the Shire than with the Clydesdale. He was a chesnut horse of great volume and weight, and, although possessed of a magnificent top, a little short and steep in his fore-pasterns, and with a narrow formation of head. He travelled in several widely divergent districts of Scotland, and for several seasons in Yorkshire, but as far as Scotland is concerned, the best of his progeny as a whole were those bred in Dumbartonshire, out of old-fashioned low-set, broad Clydesdale mares. He gained numerous prizes, including first at the Highland and Agricultural Society's Show at Stirling in 1873.

Crown Prince 207 was in many respects a first-rate type of the Clydesdale. He had good feet and legs, and a really first-class Clydesdale formation of head; *i.e.*, it was of a fair length, broad between the eyes, and altogether gave the impression of considerable intelligence. Like Old Times, he was rather long and easy in his back, and, perhaps, not too well-sprung or barrel-shaped in his rib. He was bred in Renfrewshire, and his breeding was a combination of the same blood as Samson 741, and the blood of Clyde 155, and Barr's Prince Royal 647, with a foundation of Kintyre blood. He was twice second at the Highland and Agricultural Society's Shows, and a prize winner elsewhere. His stock were not particularly distinguished in the male line, but his female progeny have conferred many and lasting benefits on the breed. They are in the main marked by the Clydesdale characteristics which have been referred to as prominent in himself, and are never lacking in substance and weight.

On the whole, the tendency in the Clydesdale breed during the past ten years has been towards greater elevation of shoulder, roundness of barrel and levelness of top—an influence easily traceable to Darnley and his tribe; and a marked improvement in action and style, as easily traceable to Prince of Wales and his tribe. Aiming at the development of qualities which are enduring rather than temporary, in some localities too much anxiety to produce fancy animals with exaggerated showyard points may have led to the neglect of more solid and enduring excellencies, while there has been a steady determination over all to keep the Clydesdale to the front. The improvement in the female line is greater than in the male line. There is a tendency to overlook the fact that

CLYDESDALE MARE, LAURA LEE.
Winner of First Prize in 1885 and 1888.
The Property of Mr. J. Jardine Peterson.

the male must be masculine, and hence a few stallions have appeared which would have secured much greater distinction had they been of the opposite sex. Auction sales have become much more numerous than formerly, and the prices realised, especially for females, have never been surpassed or before approached. Abnormally high prices have been paid for males, but these transactions have all taken place privately.

A feature of the past ten years has been the opening up of new markets in different parts of the world, and the almost complete cessation of trade with Australia and New Zealand, which in former days were the best Clydesdale markets. The chief of the new markets have been the United States, Canada, the Argentine Republic, Chili, Brazil, Germany, and Sweden. A limited number of horses have also been exported to the Cape of Good Hope.

Points of the Clydesdale.

It is not easy to find language which will adequately convey an idea of the present standard of points in the Clydesdale, mainly because the terms employed are of necessity relative, and have different shades of meaning according to the example of the breed present to the mind's eye of the writer or reader.

The old school of Clydesdale judges—that is, the school of twenty-five years ago, began to judge at the head, travelled over the back and quarters, finishing up with the limbs and feet. The new school, which began to assert itself say about fifteen years ago, begin with the feet—" no foot, no horse "—and travel upwards. We follow their example. The ideal horse of modern days in feet and limbs is Prince of Albion 6178. He was first at the Highland and Agricultural Society's Show four years in succession, a sufficient indication of the position held by him in popular esteem. He has large, round, open feet, with particularly wide coronets, and the heels are also wide and clearly defined. His pasterns are long and set back at an angle which would be considered too acute in the Shire. His bones are wide, flat, thin, and dense. The following were the measurements of several leading sires in the month of March, 1891. The ages of each, at the date of measurement, are given, and it ought to be understood that the horses were not in show condition.

	PRINCE OF ALBION 6178, f. 25th April, 1886.	SIR EVERARD 5353, f. May, 1885.	SIRDAR 4714, f. 18th April, 1884.	CAIRNBROGIE STAMP 4274, f. 29th April, '84.	FLASHWOOD 3604, f. 13th May, 1883.
Height	16 ft. 3 in. on plates	fully 17.2	17.1½	17.0½	17.0
Girth (in low condition)	7 ft. 4 in.	8 ft. (lean)	7 ft. 10 in.	7 ft. 6½ in. (lean)	8 ft.
Weight		20½ cwt. June, 1890	19½ cwt.		20 cwt.
Arm	23½ in upper muscles. 18 in. at horn	26 in. upper muscles	23 in. at top	32 in. upper muscles	20 in. above horn.
Knee		17 in. round.			
Bone below the knee	11½ in.	11 in.	10¾ in.		11 in.
Bone below the hock	12½ in.	12 in.	12¾ in.	11½ in.	12¾ in.
Length from elbow to knee—middle of joint	19½ in.		18½ in.		
Centre of knee to centre of fetlock joint	11½ in.	11½ in.	12¾ in.	11½ in.	
Stifle to bend of hock	21½ in.	21½ in.	22 in.		
Point of hock to fetlock	14½ in.	18½ in.	13½ in.		

In approaching one, the Clydesdale should carry both feet absolutely straight and level, and the whole appearance of Prince of Albion when so viewed is regarded as about perfect. He has a wide chest and low counter, but his limbs are planted well under him, and there is no tendency to what is called being wide at the shoulder, that is, having the fore-limbs so coming out of the shoulder that the horse is compelled to walk in front somewhat after the fashion of a bull-dog. The slightest inclination to this in a Clydesdale is regarded as unpardonable. The Clydesdale has an oblique shoulder, lying well back on high withers. A ewe neck, that is, a neck which carries the crown of the head at about the same level as the top of the shoulders, is not regarded with favour, and an arching high neck, whether in male or female, is

From a Photograph by Chas. Reid, Wishaw.]

CLYDESDALE STALLION, LAWRENCE CHIEF 7910.
Winner of Glasgow Agricultural Society's Premium, 1894.
The Property of Mr. Alexander Scott.

always an attraction. The head should be of medium length and broad between the eyes and at the muzzle. A tendency to "dish-face" may be observed in some tribes, and this is generally accompanied by a small ear, and what, in the main, is characterised as a "pony head." Wherever this style predominates there is probably a strain of Highland or old Galloway in the blood. On the other hand, the hard, narrow face and Roman nose are regarded as equally, if not more objectionable. Such features are usually indicative of a strain of Shire blood, and, indeed, they are not otherwise to be accounted for in the Clydesdale. An open, level countenance, vigorous eye, and large ear, are greatly valued, and not readily sacrificed. In respect of the head, neck, shoulder, back, ribs and loins, perhaps the best made Clydesdale stallion of recent years was Flashwood 3604. But the measurements already quoted will have indicated fairly well the relative measurements in important points of some of the best stallions seen during the past ten years. The hind limbs of the Clydesdale have not nearly so much attention paid to them as the fore legs—and in this, we think, Clydesdale judges err. Especially in regard to entire horses is it true that no part of their anatomy should be more carefully attended to, and broad bones, of the texture indicated as essential in the fore legs, broad, clean, sharply defined hocks, with the hams coming well down into the thighs, and the latter maintaining their strength and muscular development right down almost to the hocks, should be more insisted on than they are. The truth is that we are disposed to regard weakness in the thighs as the most undesirable blemish on the Clydesdale at the present day. If Prince of Wales 673 gave us rather more of the hard, narrow head with Roman nose, and the straight hock, than was desirable, Darnley 222 gave us too great a lack of muscular development in the thighs, and rather a sudden droop in the quarters. Thoroughbred quarters are not asked for in the Clydesdale, but, on the other hand, neither are the quarters of the Percheron. The tail should be well set on, by which we mean that it should be set well up, and the quarters and thighs should not be too sharply marked off.

Action is all important in the Clydesdale. Even his most severe critic will not deny that in this particular he generally excels. He is never judged travelling round about the ring, but always up and down the centre in front of his judges. Hence his limbs must be squarely planted under him; they must follow each other in an undeviating line, and and it is an all-important requisite that the points of the hocks be inclined inward and not outward. A Clydesdale must stand with its hind legs in regulation military form—heels in and toes out. Any other arrangement is tabooed, and if perfection is not always attained, it is always sought for, and many things are sacrificed to secure the prize for an animal which keeps its hocks well together. The consequence of the attention bestowed on action is that the Clydesdale as a rule is both a good walker and a good trotter. Some of the best show horses have had trotting action almost equal to that of the best Hackneys, while there probably never was a stallion of any breed which could have excelled Darnley at the walking pace. The chief improvements effected in the Clydesdale during the past twelve years are in our opinion these: An increase in the quality, by which we mean the density, and wearing properties of the bones; a marked advance in the direction of deepening the rib, shortening the coupling, and rounding the barrel; a gradual but quite discernible return to the old Clydesdale type of head, and a very distinct advance in general soundness and freedom from the diseases scheduled as hereditary unsoundnesses by the Royal College of Veterinary Surgeons. In proof of this it is but necessary to refer to the results of the veterinary examinations at the Royal Agricultural Society's Shows since these came into force four or five years ago. None of the other draught breeds has come anything like so well through this ordeal as the Clydesdale. As regards the popular size of the Clydesdale, the figures already quoted

From a Photograph by Chas. Reid.]   CLYDESDALE MARE, SUNRISE.
Winner of First and Champion Prize at R.A.S.E., Windsor, 1889; First at H & A. S.
Shows, 1889 and 1890.
The Property of Mr. David Riddell.

relative to leading stallions are the best testimony. Mares as a rule are about an inch less in height than the stallions, and their other measurements are in proportion. It is, however, worthy of remark that there are many Clydesdale mares of quite exceptional size and weight—such for example is the Marquis of Londonderry's Primula 7477, which in a large class of dray horses won first prize at a recent Durham County Show. That there has been a tendency in the show-yard during the past ten years, to favour "bonnie" animals rather than strong animals, is not to be denied, but during the past season, and even more emphatically during the present season (1894) judges have abandoned this fancy, and the draught horse type is decidedly in the ascendant.

CHAPTER IV.

THE BREEDING OF HEAVY CART HORSES FOR STREET WORK.

[It may be mentioned here that the writer of this chapter is Mr. W. R. Trotter].

FEW people will deny that the breeding of cart horses for street work is one of the most important departments of British farming at the present time. The prices of farm produce have come down and down, despite the prognostications of the most learned, and at the time of writing, flour of the very best quality is being sold at 1s. per imperial stone, the best beef at 6d. per lb., and other products at a similar low rate. This, added to a disastrous season like 1893, when the greater part of England suffered from an exceptional drought, is calculated to make all thoughtful men turn their attention to the production of a description of stock which has not yet been seriously threatened by foreign competition. At the present time £100 for a seasoned heavy cart gelding is not an uncommon price; several have been sold at much more, and the general run of prices for heavy horses is from £70 to £100. This forms a striking contrast to the value of other produce which the farmer has to sell.

It will be impossible to give such advice as will insure universal success in the production of this class of stock, as horses are, of course, susceptible to the influences of soil and climate. The treatment that is practised on one farm is not always applicable to another holding, and all men do not manage alike.

The title of this chapter will indicate that I do not mean to enter upon an elaborate history of the cart horses of the British Isles, as that subject has been discussed elsewhere. It is more the intention to take an impartial view of matters as they are, and to give some hints at improving them, seeing that "there is no time like the present." However, it would not be wise to wholly ignore all modern history, for we have had several prominent examples of what can be done by judicious care and judgment in cart horse breeding as well as in other things. There is no doubt that the heavy horses of England and Scotland are of one and the same family, and that an interchanging of breeding stock has been going on over the Border for the whole of this century, at any rate. I do not, of course, refer to the depredations of the Border clans, who thought it part of their duty to cross the Border and carry back as much of their enemies' stock as they could lay hands upon. According to several old horse books that have come under the notice of the writer, English dealers early this century travelled north to Rutherglen fair, and bought large lots of cart fillies, which were taken into Lancashire and several other counties. How long this trade continued it is not easy to define, but certain it is that in the neighbourhood of Carlisle, Longtown, and several other Border towns, tribes of horse dealers now exist, and have existed, in the same families for generations, whose operations in their trade have been pursued from Preston in the South to Glasgow, Falkirk, and Edinburgh in the North. When they found they could dispose of good fillies in the North they often drew their supplies from the South, and *vice versâ*. The direction of the current undoubtedly varied, but according to available history it was certainly from North to South during the early half of this century. It is only during the last thirty years that the Scottish invasion proper commenced. The Scotch dealers began to find that any number of beautiful fillies of the Clydesdale type could be procured in the Midlands. Derbyshire, Nottinghamshire, Lincolnshire, and Cambridgeshire received the most of their attentions. Welshpool fair was regularly visited; in fact, one of the best geldings known to the writer to-day was got by a stallion whose dam was bought at Welshpool, and whose sire was a Clydesdale. The men who principally took such a lot of grand mares North were the late Hugh Crawford, David Riddell, and many others. On one occasion Mr. Riddell bought as many as forty fillies at Waltham fair, as bonnie Clydesdales as ever were seen, according to his opinion (which is always worth having about a cart horse). Mr. Drew drove for weeks round the nooks and corners of Derbyshire, often piloted by Mr. Samuel Wade, of Mickleover, who used to tell how Mr. Drew would travel down by the night train from Glasgow to Derby, and drive out to Mr. Wade's place, where he would arrive about four o'clock in the morning, and immediately commence to disturb the slumbers of Mr. Wade by throwing pebbles up to the window, a plan that has been adopted by many another man bent on a wooing expedition. On one occasion these two gentlemen secured fourteen choice fillies, nearly all by Lincolnshire Lad, which, of course, Mr. Wade expected would satisfy the cravings of the shrewd Scotchman for some time, but much to his surprise the genial tenant of Merryton returned in about a fortnight for a fresh lot of his favourite sort. This went on for several years, and in the early half of the seventies the bulk of the mares and fillies disposed of at Mr. Drew's sensational sales hailed from the Midlands, and these were diffused over the length and breadth of Scotland. The late Mr. Hugh Crawford was once met by a friend in Carlisle station, and on being asked where he had been, said he had been to England for a "wheen horse" (a few horses), but the few turned out to be a whole special train load. These facts may give some idea of the magnitude of the trade, and may account in a measure for the great difficulty there is in England to find mares with correct ankles, good feet, and clean coronets, because for years the Scotchman took away only those animals which excelled in these cardinal points, thus leaving a residuum of short pasterned animals, with defective feet, and having of side-bones above an average crop.

At this time there was a sort of reciprocity going on in the opposite direction. Many English noblemen employed Scotch bailiffs and managers, and these men often succeeded in taking some Clydesdale horses with them. The great Clydesdale horse Lofty, or Young Lofty 987 was taken into Gloucestershire. He subsequently came into Derbyshire, where he travelled for years, and was known as Tagg's Lofty. Scores of grand horses were produced from him in the Burton district, notably Drew's Countess, that won first prize at the "Royal" show four times, and the highest award at the Paris Exhibition. Her full sister was White's Farmer, that gained the champion prize at Ashbourne show when nearly 20 years old. This mare bred well, having produced Lord Ellesmere's Farmer, probably the best mare that was ever at Worsley, and she, again, is the dam of Mr. Salt's William the Conqueror horse, Duke of Normandy, the sire of Mr. Muntz's Willington Boy, who, by the by, has Lofty for great grandsire on his dam's side, so he has a double cross of the Clydesdale, and, in addition, he is of the same colour as Countess and her sister. White's Farmer also produced the great Royal Albert mare, Pauline, that was sold to the late Mr. Punchard as a filly for 300 guineas. Sir Walter Scott, the grandsire of the celebrated Prince of Wales, travelled in the Fylde district in Lancashire for several seasons until his great merits as a breeder were discovered by Mr. Riddell, who at once tracked him, and took him back to his native land, where it is needless to say his descendants have effected a revolution. For this one act alone the Clydesdale world should be for ever grateful to Mr. Riddell. Mr. Thomas Shaw, of Winmarleigh, kept Clydesdale stallions for many years, and his old horse, Argyle, whose blood runs in many of the best horses in Lancashire to-day, was only four or five years ago standing near Carnforth, in North Lancashire. Mr. T. H. Miller's celebrated Princess Dagmar was a direct descendant of Tom o' the Gills, a Clydesdale horse bred in Cumberland. Mr. Charles W. Tindall took a Clydesdale horse to Mr. John Torr's, in Lincolnshire, in 1875, and he has said that a Clydesdale horse named Ronald McDonald left a lot of grand geldings off Lincolnshire-bred mares. The Duke of Richmond also introduced Clydesdales to Goodwood, and they are still kept there. Mr. Stewart Hodgson had a nice stud in Surrey, where they did a lot of good, and the Lords Cecil have at present a fine lot at Tunbridge, in Kent; so also has Lord Cawdor in South Wales. The Duke of Portland for many years had a show team of Clydesdales at Welbeck, and for several seasons they competed with Shires, and won a large number of prizes.

Sufficient has probably been said of a historical nature to show us that the breeds have been closely interwoven for many generations. It will be necessary later on to examine more closely the details of the breeding of some of the most prominent horses of the day in order to prove exactly how they are bred, so that it may be a guide to us in the production of similar high class horses that can make such remunerative prices for town work. It is, however, rather remarkable that the Scotch breeders have always made good feet and correct pasterns leading points in their operations, and for a long time their English brethren paid little or no attention to these essential points, and bestowed their principal attention to weight and formation of body, so that to take the majority of Scotch horses to-day, they show a far greater uniformity of type than the English, with the most beautiful of feet and ankles, and fine quality of bone. Their greatest deficiency is weight of body—often they are rather light in ribs, and they could do with more bone; while the rank and file of the English branch is certainly thoroughly superior where the Scotch is inferior, and *vice versâ*. Of course, it is not meant that individual specimens cannot easily be found possessing all the good points of both breeds, but speaking generally these remarks hold good.

It will be readily gathered from the foregoing that in the opinion of the writer, a sounder, better and more saleable gelding for town work can be produced in the most expeditious way by a judicious blending of Clydesdale and Shire blood than by sticking closely to Stud Book lines. The wisdom of this course is more clearly demonstrated in Scotland, where, as previously stated, an enormous number of English mares have been taken and put to Clydesdale horses. The produce inherit from their Shire dams far more substance of body than was possessed by the old original longbodied Clydesdale mares, while the correctness of formation of limbs, clean coronets and big, sound, well-shaped feet are transmitted from their Clydesdale sires. These latter highly essential points have been cultivated and thoroughly developed by years of careful breeding and by rejecting the stallions with faulty limbs and defective feet. To illustrate this statement I may mention several noted stallions bred or owned by the late Mr. Lawrence Drew, of Merryton, who was unquestionably the pioneer in this system of breeding cart horses, and certain it is that no one in modern history achieved such distinguished success as a breeder. In this respect it may truly be said that Mr. Drew was a national benefactor. Among the horses he bred, or that were bred from the English mares he took North, we may mention those celebrated animals, Lord Harry, Prince Imperial, Lord Douglas (Glasgow winner) Roderick Dhu (Glasgow winner) and Rosebery (who was second at the same show), Duke of Hamilton (sold for 1,000 guineas), Prince George of Wales, Brilliant, Luck's All, St. Lawrence (Glasgow winner), St. Mungo, St. Vincent, Hawkhead (Glasgow winner), Prince of Avondale (twice Glasgow winner), Pearl of Avondale, Bold Briton, Premier Prince, Brave Wallace, Clarendon, Bonnie Prince, Mains of

Airies, Handsome Prince, Gallant Prince, &c., &c. Some of these horses were eligible for the Clydesdale Stud Book, although they all inherited both Shire and Clydesdale blood. To show the distinction some of them attained I will take as specimens Prince of Avondale, Prince Lawrence, Castlereagh and Lord Ailsa, all of which were descended from Mr. Drew's English mares.

(1) Prince of Avondale, foaled in 1880, sire Prince of Wales, dam Juno, by the Shire stallion Ploughboy 1741. This horse was bred from a mare bought by Mr. Drew at Waltham-on-the-Wold fair. I think she came through the hands of Mr. H. Freshney. As a yearling he gained first prizes at Edinburgh, Strathaven, Kilbride, and the Highland and Agricultural Society's Shows; he also won the first prize at the Glasgow Stallion Show in 1883, and again in 1884, and travelled the district. He will be best remembered by English readers as winning the first prize at the Royal Agricultural Society's Show at Norwich. He was not very extensively patronised in the Glasgow district, owing to his not being registered in the Stud Book, and unfortunately he proved a very unprolific sire. What he did leave were, however, of exceptional excellence, and it is quite certain that no other stallion in Scotland has produced such a pair of superb mares as Rose of Banknock and Sunray. Rose of Banknock won innumerable prizes previous to becoming the property of Mr. A. H. Boyle, of Banknock, and her record was without precedent. In 1889 she commenced at Kilmarnock by winning first as a brood mare, the silver medal as the best female, and the Duke of Portland's cup for the best of the breed; at Ayr, first prize in her class; at Maryhill, first prize in her class, and special as best female on the ground; at Glasgow, first prize in her class, and silver cup for best of the breed; at East Kilbride, first in her class, and silver cup as best of the breed; she was also awarded similar honours at the Falkirk, the Kirkintilloch, the Lanark, the Dunblane and Edinburgh Shows; she also won the first

prize at the Melrose Show of the Highland and Agricultural Society, and the cup as the best female exhibited. It will thus be observed that she not only won every time she was exhibited, but in addition she gained the special prizes (where offered) as the best of her breed, or the best in the yard. Such a record is unparalleled, and needs no comment. The other celebrated daughter of Prince of Avondale, Sunray, is owned by Mr. Mitchell, Polmont. She won numerous prizes when a filly; as a brood mare she was first at Ayr; first at Maryhill and champion medal; first at Barrhead and champion medal; first at Hamilton and champion medal; first at East Kilbride and champion cup; first at Springburn and medal; first Edinburgh (her foal also first); first Highland and Agricultural Society's Show, Glasgow. In 1889 she slipped her foal, and was not in good show form, but won first at Maryhill, first at Glasgow (where she beat Ayr and Kilmarnock winners), first Hamilton, first Springburn and first at Kirkintilloch (she was not shown at any other show). It will thus be apparent that not only did she herself win the highest honours, but that she produced an exceptionally good foal that won first prize on several occasions. She is also the dam of the unbeaten three-year old colt Prince of Millfield. Prince of Avondale is also the sire of Sir James Duke's beautiful horse Fashion, that was second at the Highland and Agricultural Show at Glasgow; also of Mr. David Riddell's good-looking Golden Avon, one of the horses in the short leet of three-year-olds at Glasgow, and winner at the Highland and Agricultural Society's Melrose Show.

(2) Prince Lawrence (lately owned by Mr. Peter Crawford, Eastfield House, Dumfries), is not eligible for the Clydesdale Stud Book, but the Clydesdale Horse Society have, by a recent rule, shown a wise discretion in admitting his produce from registered mares. No horse has made such a mark in such a short time in modern history, and it is a national misfortune that he should have died so soon. As to his breeding, he is in-bred to the veteran Prince of Wales,

being got by Prince George of Wales (by Prince of Wales), dam by Prince David (by Prince of Wales). His sire, Prince George of Wales, was bred by Mr. Drew, and was out of his well-known English mare Jessie Brown, and she was believed to be got by Bold Lincoln 231; consequently Prince Lawrence is a grandson of an English mare; he has also a concentration of the blood of Prince of Wales, himself half English. As a three-year-old he was second at Glasgow Stallion Show, and was first the following year at the Highland and Agricultural Society's meeting at Perth. Since then he has not been exhibited, but his career at the stud has been phenomenal, as the following will show. For his first two seasons he travelled at moderate fees, but in 1888 he travelled the Girvan district, terms £3 at service, and £4 each foal. In 1889 he travelled for the Brechin and Perth Horse Club at £3 and £5. He was let to the same Society for 1890 at £10 and £3, and for season 1890 terms of £10 and £10 were refused from a different society. His first crop of foals in Glenkins district were of great excellence, and contained Lady Lawrence, sold to Lord Cawdor for £400; Eastfield Chief and Eastfield Model, sold for £1,000; Lawrence's Heir, Eastfield Laird, &c., &c. In the prize ring as a lot they stood practically undefeated. In family competitions for the best five two-year-olds by one horse they were first at Kilmarnock Show and first at Glasgow Show, also second for five yearlings by one horse at the same show. The most prominent winner in the two-year-olds was Lady Lawrence; as a yearling she was first at Kilmarnock (Derby), second at Ayr, first Maryhill, first Hamilton, first East Kilbride; as a two-year-old, first at the Royal Show at Windsor (fifteen shown), first Hamilton and champion cup, second at Kilmarnock. At Kilmarnock Show Lawrence's Heir was second, and Eastfield Chief fourth; Ayr Show, Eastfield Chief second; Glasgow Show, Lawrence's Heir second, Eastfield Chief third, Eastfield Model sixth, Eastfield Laird commended; Edinburgh Show, Eastfield Chief first. Provincial fair, Canada, Eastfield Chief first; also first at the Industrial Show. Lawrence's Heir was let to the Perth, Glamis, and Brechin Society in the place of his sire for 1890; terms, £4 at service, and £6 for every foal. Prince Lawrence's yearlings were second at Kilmarnock, first at Glasgow, first Edinburgh, first and fourth Greenock, second Highland and Agricultural Society, Melrose, first Barrhead Open Show, second Paisley, first Arbroath, and champion as best female, and second Royal at Windsor. A foal by Prince Lawrence was sold to Mr. A. Scott, Greenock, for £375 net. It would be difficult to find a horse that has made such a record; his first crop of foals were invincible as a group of two-year-olds, in addition to winning prizes individually at the leading shows of Clydesdales; his second crop also distinguished themselves in a similar manner, and the price above noticed for a foal is a sufficient proof of the great excellence of his third crop. The death of Prince Lawrence was not only a great loss to his owner, but to the Clydesdale breed it was incalculable.

(3) Castlereagh, owned by the Marquis of Londonderry, is probably the most familiar horse to English readers. He will be remembered as winning the first prize at the Royal Show at Newcastle in 1887, also first at both the Great Yorkshire Show at Huddersfield and at the Lancashire Show at Lancaster in 1888. Many people thought him very badly used at the Perth Show of the Highland and Agricultural Society, where some prominent judges put him down an easy winner, and again at Melrose he might have been second to Lord Ailsa, and no injustice done to any one. He was certainly the most typical Clydesdale horse in the ring, and a well-known American declared him by far the most suitable and valuable horse for his country if he had been registered. In breeding he may be termed the orthodox cross on the top, being by Darnley, dam by Prince of Wales; his granddam was Mr. Drew's black Shire mare Topsy, that will be remembered as winning first prize in her class, and champion as the

best female at the Shire Horse Show in London in 1880. She was bred in Derbyshire, and bought by Mr. Drew from Mr. Goodall, Milton, Derbyshire; being got by Crown Prince 558, dam by William the Conqueror 2,343. (Crown Prince was from a mare by William the Conqueror 2,340, the grandsire of the great horse of that name numbered 2343.) It will thus be seen that Topsy was closely related and in-bred to one of the most celebrated Shire horses of the day William the Conqueror 2343, the sire of Prince William, Staunton Hero, Hitchin Conqueror, Electric, and Endymion. It would therefore be difficult to imagine a horse better bred than Castlereagh, combining as he does the blood of the two greatest horses Scotland has ever seen, and the best blood in the whole Shire Horse Stud Book. Although Castlereagh is not entered in the Clydesdale Stud Book, all his produce from registered mares are eligible. As may be imagined from such breeding, Castlereagh is an exceptionally good getter. He has been kept almost exclusively for use in Lord Londonderry's own stud, and an inspection of that stud will confirm this. His produce have won a great number of prizes in Northumberland and Durham, and also at the Royal and Highland Society's Shows during recent years. The following are a few of the honours won by them in 1889:—Loyalist 6022 won first prize at Hamilton Show, and was also awarded the Lesmahagow premium to travel their district. Winnie, first prize yearling filly at Kilbride Open Show, and second at Paisley. Yearling filly, dam Cowslip, first at Northumberland Show, first at Durham County Show, and second at Highland Society's Show at Melrose. Yearling colt, first at both the Northumberland and Durham County Shows, and medals of the Clydesdale Horse Society. Two-year-old filly Gladys, *vhc* at Royal Windsor Show, second Durham County Show, first Northumberland Show. Rowan, three-year-old filly, first Darlington Show, first at Brampton Show. The Cumberland Agricultural Society offer prizes at Carlisle for the best pair of fillies under three years old. One year Mr. Sinclair Scott sent his pair (the best in Scotland) and won; the next year several celebrities from over the Border were present, but the first and second prizes were won by four fillies, all by Castlereagh, viz., Lauristina and Letitia first, Rowan and Gladys second. His stock also obtained leading honours at the Cleveland, Northallerton, and Darlington Shows in England, and at Stranraer and Ayr in Scotland.

(4) Lord Ailsa, owned by Mr. John Galbraith, Croy Cunningham. This horse is entered in the Stud Book as got by Lord Erskine, dam Jewel, by Prince of Wales; the breeding of his dam was subsequently challenged, and some people declared her to be a pure English mare. The full particulars of these statements were published in the *Live Stock Journal*. However, the subject was fully gone into by the Clydesdale Horse Society, and the registration confirmed. The contention of the parties who challenged the pedigree was that Jewel, when sold at Mr. Drew's sale, was entered in the catalogue, no sire being given, and that when Mr. Drew was asked the question he did not say her sire was Prince of Wales. A perusal of Mr. Drew's catalogues will not show a single case in which, where an animal was got by Prince of Wales, the fact was not stated, and in every catalogue (except the one issued as a register by Mr. Drew and the one published after his death) the English horses are entered without pedigrees. Let the horse be bred as he may, he is certainly a credit to his ancestors. It is, of course, conceded by everybody that his granddam was English. His success at the Highland Show at Melrose was almost universally approved of. He is a horse of great size and marvellous quality. A very beautiful and good likeness of him appeared in the Chicago *Breeders' Gazette*. He had the Bute premium of £100 in 1888, and his foals were not only very plentiful, but were of such uniformity and excellence that Mr. Galbraith bought twelve of them. In 1889 he was awarded the Strathendrick premium of £100. He was let to the same Society

for 1890; terms, £100 prize, and 100 mares guaranteed at £3 and £4 10s. for every foal.

It may be asked what proof there is that these horses can get good geldings? There is probably no county in England, according to its cultivable area, that produces such a large number of valuable heavy cart geldings as Cumberland. In that county both Shire and Clydesdale stallions have travelled, and most of the breeding animals contain the blood of both breeds; the stallions now being used, although mostly in the Clydesdale Stud Book, are large, powerful horses, with big and correct limbs and grand feet. Probably the best breeding horse of the lot was Lord Lothian, whose produce, year after year, obtained a host of prizes in the gelding competitions in the county; this horse's sire and grandsire have a strong admixture of Shire blood. There is no doubt that the soil, pasturage and climate of Cumberland are specially adapted for the growth of these horses. The results are highly satisfactory from a financial point of view to the farmers who breed them, and a credit to them for the way they are brought to the various sales where they are annually disposed of. Speaking of the influences of soil and climate upon the growth of carthorses, Mr. David Riddell is of the opinion that there is a marked difference in localities. He says, "It is my opinion that Clydesdales bred in Lincolnshire get much grosser and stronger. They are up to more weight, I should say by 2 cwt. each animal, than those bred in our county. I have had the practical proof of knowing this."

Physiology of Breeding.

There is not the slightest doubt that a careful study of the physiology of breeding is of the very greatest importance in the consideration of cart-horse breeding. About fifty years ago a medical practitioner, named Dr. Orton, a resident in Sunderland, after carefully experimenting for several years, propounded the theory that in breeding animals, in the majority of cases the male parent influenced in a great measure the outward formation, principally the external structure or locomotive organs; and the female the internal organisation, viz., the whole circulatory, respiratory and digestive organs.

A number of years later, in the Journal of the Royal Agricultural Society of England, published in 1865, Mr. W. C. Spooner contributed a most interesting article on the cross breeding of horses, and advanced the same theory as that of Dr. Orton, with very slight variation. He was, however, of opinion, that the female often gives the head and neck, and the male the back and hind quarters. Mr. Spooner was undoubtedly a very careful observer, and he adduced a large number of instances that came within his own knowledge, where the theory worked out with striking correctness, proving most conclusively that in breeding horses, as well as other live stock, we can reduce the uncertainties to a very considerable extent. The writer is quite satisfied from careful observations in the breeding of cart horses in his own stud that the locomotive organs of the sire are far oftener transmitted than those of the dam. He can trace the exact walking and trotting action transmitted through four generations on the sire's side; the fourth generation now at the stud is producing foals of the almost identical type to himself, particularly in the hind quarters, locomotive organs and action. The famous Clydesdale stallion Darnley, probably one of the greatest horses of the century, was a striking illustration of the transmission of faulty hind quarters by the sire. This horse, although certainly good enough over his rumps and tail-head, was undoubtedly deficient below his tail, being decidedly light thighed. Nearly the whole of his produce inherit this defect in a greater or less degree. Many other striking illustrations could be brought forward where stallions that were narrow and split up behind transmitted that defect with decidedly unpleasant persistency. The late Mr. James Howard, of Bedford, who had studied breeding from Dr. Orton's and Mr. Spooner's standpoint, in referring thereto, said: "Observation and experience have satisfied me that they are sound, and although like every other breeder I know something of the uncertainties of breeding, yet I am convinced that there are certain laws pertaining to the process which cannot be disregarded with impunity." Mr. Drew undoubtedly acted as if he was thoroughly conversant with this theory. His great horse, Prince of Wales, was certainly very deficient in ribs, but his limbs were extraordinary for quality and formation—beautiful pasterns and flexible coronets, free from side-bones, open hoof heads and perfect feet, and his action was superb. Mr. Drew, as has been previously stated, obtained his supplies of thick-bodied Shire mares from the Midlands, with, of course, the best feet and ankles he could get, and the result was a perfect confirmation of the theories here advanced, because the produce inherited in a marked degree the limbs and action of the old Prince, and the thick bodies of their dams. His record as a breeder stands unrivalled at the present day. It will probably repay cart-horse breeders to cast some retrospective thoughts on the various stallions and mares they have known in the past, and see if they cannot trace the influence of the respective parents in the produce of those animals, and thus satisfy themselves on this most important point; this would certainly be a great guide to their future procedure. Probably some of the horses mentioned hereafter may be known to them. The following will show that the most successful breeding horses in modern times have been those with the best of limbs or locomotive organs, though many of them were deficient in ribs.

William the Conqueror, the sire of the three London champions, Prince William, Hitchin Conqueror, and Staunton Hero, was decidedly a light bodied horse. When at Worsley, over 20 years old, his legs, feet and pasterns were nearly perfect; in fact there are few stallions of any age to-day with the same superior set of limbs, and the way he moved them was a treat to see and almost a greater treat to hear. The elasticity that can only exist in perfectly shaped limbs was unimpaired by twenty years of stud life.

Royal Albert was a very tall horse, with the biggest of hind legs, broad, flat bones and ample pasterns. His son Albert Edward is a similar horse and a good breeder. When Royal Albert served in his native district, where his ancestors for several generations had travelled, and where the mares were rather on the leg like himself, he left certainly a lot of useful horses, and several rather tall ones, but it was not until he migrated among the thick-bodied, short-legged mares in the Ashbourne district that he became the sire of such high-class produce.

Bar None, when bought as a three-year-old, was leggy and narrow; yet this horse was the best four-year-old stallion of his year, and is probably the sire of more good mares than any other stallion; but his limbs were superb, and this is what he imparted to his progeny. Lincolnshire Lad, and his two sons—Lincolnshire Lad 2nd and Hydraulic—were all leggy horses, but their limbs were of great size and quality. Mr. Drew wrote in 1881, that the old horse was the sire of more prize mares at the principal shows in Scotland during the previous five years than any other horse in the country. Lincolnshire Lad 2nd, sire of Harold, &c., has more bone and hair than any other horse in England, but is light in his ribs and narrow in front. Hitchin Conqueror, one of the best all-round getters in England to-day, is decidedly a tall horse, and for some years was badly used in London on this account. Sir Colin, the sire of Starlight, and of many grand brood mares in Lancashire, including the dam of Vulcan (twice London Champion), was an old horse when at Worsley, but decidedly on the leg. Premier and his sire, What's Wanted, were both light-middled horses with great limbs. Among Clydesdales, the greatest breeding stallion I have ever known, Prince of Wales, has been previously referred to. Darnley, in the eyes of English judges, would have been considered a leggy horse, and Lord Lyon was certainly in the same category. Mr. Peter Crawford's Prince Lawrence, the most successful breeding young horse of recent years, was so leggy and narrow when he won first prize at the Highland and Agricultural Show at Perth, that many people complained of the decision, but his limbs were so remarkable for size, quality and formation, that the judges could not get over him; and, as we have already stated, subsequent events proved that they were right.

Before proceeding to consider the selection of breeding animals, probably it will be desirable to review the most important feature of the whole subject, viz., soundness. Undoubtedly, the large preponderance of unsound material is the greatest drawback to the successful production of marketable animals. It is admitted that the following diseases are notoriously hereditary, viz., side-bones, ring-bones, spavins, navicular disease, curbs, stringhalt, shivering, boggy hocks, roaring and whistling. The most prevalent of these in carthorses is side-bones and roaring; but although the others are not quite so common, their presence must be avoided if possible. An eminent authority on horse-breeding, who is also a qualified veterinary surgeon, is of opinion that the depreciation in value of a heavy cart gelding, worth, say £90 if sound, through having side-bones would be £30; but he further states that the actual working value of the animal is not so seriously deteriorated if he has an ample hoof-head (or coronet) and good strong open feet, but of course these things are not always considered by men who buy all their workhorses only on the condition that their veterinary will pass them sound and clean, and it is very doubtful if the above figures adequately represent the depreciation in the market value of a gelding. How, then, have side-bones to be got rid of? The same authority is of opinion that they cannot be bred out except at the expense of a deterioration of size and weight; certain it is that in all horses, the bigger they are the larger percentage there are of unsound ones; and as a proof of the very serious number of side-boned horses there are in the country, a circumstance which occurred at a sale of Shire horses some time ago may be mentioned. A very good foal was to be offered from a mare with no less than four side-bones (the foal was ultimately sold at a high price); a well-known horse-breeder in conversation put this question to the writer, viz., "If he thought the foal could be safely bought for a stud horse?" "Certainly not," was the reply; "the mare is not fit to breed a stallion with such unsound feet." A prominent breeder chimed in, "But where are you going to get them without side-bones?" It looked a hopeless business to attempt to advance the breeding of sound horses in such company, if such a state of things existed. If that worthy man would take a trip to Glasgow Stallion Show, he would find side-bones few and far between, and horses of sufficient size and weight for any purpose. It is not meant that he would find *all* the horses as large as he would like, because the Scotch taste is somewhat different to the English in this connection; they go more for quality than size. Then, again, there are many Shire horses shown in London, year after year, that keep clear of side-bones, and are of sufficient size. Of course it cannot be denied that they are not sufficiently plentiful to meet the demand there is for them. There is certainly no possible reason why side-bones cannot be bred out as well as other similar diseases, remembering that they are a disease of the locomotive organs; it certainly points to the strong advisability of causing all stallions with such objectionable ornaments to be castrated at once. There are strict laws for the suppression of diseases in cattle, sheep, and pigs, and it is equally necessary for Government to intervene in respect of horse-breeding. But it is quite possible to show that side-bones have been, and are being, bred out. It is well-known that a famous old Shire horse had them, as had also many of his produce; one celebrated mare by him taken into Scotland was similarly affected, yet nearly all her produce by Prince of Wales were clear, and one of her daughters is now a distinguished prize winner and the dam of two of the soundest and most successful breeding stallions in Scotland. They may be excused in a mare but never in a stallion. If only those stallions that are clear of side-bones were used, depend upon it we would soon get plenty of horses with clean coronets. It might be reasonably assumed that roaring being an affection of the wind is a secondary unsoundness to side-bones and other similar complaints which affect the locomotive organs, seeing that roaring and whistling cannot affect the actual muscular strength of the animal, and certainly cannot lame him, but men who are in the trade attach as much importance to roaring as to side-bones, and it is certainly safe to say that the decrease in value, when horses are affected in this manner, is to the same extent as a side-boned animal. It is a debateable point whether affections of the larynx and throat are more likely to be transmitted by the male or female parent, as striking instances could be adduced where they have followed both parents. But seeing that their presence seriously affects the market value of the animal, it would clearly be wise to avoid them in all breeding stock. The other forms of unsoundness are of much more rare occurrence in cart horses; at the same time their hereditary nature must not be treated lightly. The foregoing remarks as to physiology of breeding and the hereditary nature of unsoundness may be of some value as a guide to the selection of breeding stock. A careful study of the subject will certainly narrow down the uncertainty of breeding sound animals when it is conducted in the usual haphazard way, and place the subject on a firmer basis on which to operate with some degree of certainty. In the

Selection of a Stallion

we must pay the most particular care and attention to the size, quality and formation of his limbs and locomotive organs; it is not a question only of weight and width of carcase. He should, in the first place, have good-sized, sound, open feet, not abnormally large feet, but hollow below, with strong heels, and thick, tough crusts. Recently, Professor McCall clearly demonstrated that big, flat, overgrown feet were often weak and the first place where a heavy horse would go wrong, if overtaken by any serious illness; but as many mares have small feet, defective in formation, it must be a leading point that the feet of a stallion should be of sufficient size and strength, and perfect in formation; open coronets and sloping pasterns are equally requisite. So also are big knees and hocks, good quality of bone and fully developed tendons. Too much importance cannot be attached to the full development of the tendons and ligaments; they must fill the hand and be well away from the bone. A horse with weak and badly developed tendons, stuck close up against his cannon bones, always measures badly below his knee, and consequently is very liable to suffer from sprains and contraction of the back tendons. A stallion should also have strong, muscular arms and thighs, and powerful, wide quarters; to put it shortly, he ought to have plenty of propelling power behind. Action is highly important in a stallion, and undoubtedly is very likely to be hereditary, especially the walking pace, the most important gait. William the Conqueror was a fine walker and Royal Albert a bad one, and it is astonishing to find how the bulk of their produce take after them in this respect. It must not be inferred that substance in a stallion should be overlooked, as it is of considerable value, but a stallion with the best of limbs, though lacking substance of barrel, is much to be preferred to a big bodied horse with round, defective limbs and moderate feet. It is somewhat strange that out of the enormous number of stallions used in the United Kingdom every year there is such a small proportion of really good and reliable getters. In all breeds of horses we find a few stallions whose produce stand head and shoulders above all their compeers; it is needless to go into fuller details, but all who have carefully watched horse breeding will readily admit the fact; but it almost invariably happens that these horses are of considerable showyard merit themselves, or are the immediate descendants of animals of distinguished individual excellence. It will be readily conceded that, after all, a stallion's value rests in his good getting capabilities; therefore a horse that has proved himself a good getter is to be preferred to a horse untried at the stud. It is a remarkable fact that nearly all distinguished breeding stallions have had exceptionally good mothers. The dams of Darnley and Prince of Wales were the two best mares in Scotland in their day. Several good judges assert that Darnley's mother was the best mare they ever saw. Hitchin Conqueror's mother is a really good mare. In Hackneys, the dam of Denmark was a marvellous mare, and won the first prize at the Yorkshire show when considerably over twenty years of age. In thoroughbreds, Beeswing (the dam of Newminster, whose blood flows in the veins of nearly all the best horses of the day) was a wonderful mare, and won more Queen's plates and cups than any other thoroughbred mare known to history. It is certainly highly essential that an untried stallion should be from a good dam.

Brood Mares.

In these times of extreme agricultural depression it is idle to go round and tell farmers, who have not quite as good mares as they should have, to go and buy better ones. At the same time there are often very well-bred mares rather undersized, but with plenty of substance that can be bought worth the money. A brood mare should be well-ribbed and wide, with length, depth, ample heart room, and a robust constitution. And remembering that the produce often take after their dams in stamina and staying, it is highly necessary to have mares possessing these qualifications. Light-ribbed, fretty, tearing mares, are unsatisfactory to work, and often unsuitable to breed from. What may be considered undersized mares often breed well, if they possess symmetry, quality and substance. Mr. R. S. Reynolds, of Liverpool, in his essay on Cart-horse Breeding, says that the three best geldings he ever saw were from a little Welsh mare, about 15 hands 2 ins. high. Above all, breed from sound mares, if possible; a stallion cannot do all, be he ever so good. It gives a stallion a poor chance if unsound and weedy mares are put to him. Farmers are often tempted by the offer of high price to sell their good mares; if they wish to succeed in breeding more good ones they should keep them.

JUDICIOUS MATING.

Breeding is undoubtedly an art or a natural gift. We see men who distinguish themselves as horse breeders as well as in other walks of life. To be successful it is highly necessary to weigh up the respective merits of mares and stallions, so that the defects in either parent may be modified or rectified by the strong points in the opposite parent, or, to quote the words of Sir Walter Scott, "They are blended into harmony." Animals will be most likely to transmit to their progeny their very marked peculiarities, defects or strong points; and it is also certain that the more pronounced and developed any unsoundness or defect becomes the more hereditary it is. Of course, the description of desirable breeding animals hereincontained, must not be taken as the only road to success, as there must necessarily be exceptions to all rules, more particularly in breeding animals; and if a farmer has a mare rather on the leg, it is quite possible to breed successfully from her with a short-legged, thick-bodied stallion, but the paramount importance of having a stallion with really good legs must not be lost sight of.

PEDIGREE.

The value of good breeding or pedigree is highly important, for, as a rule, the more thorough and complete the concentration of good blood in any animal the more impressive he will be; but good pedigree in an inferior animal is seldom valuable. Length of pedigree is certainly valuable if its component parts possessed high individual merit. Greater faith can be placed in breeding from an animal whose immediate ancestors were known to have been high-class animals, than from one whose pedigree was long, but with nothing exceptionally good on the top. The purity of an animal is proved by the transmission of its distinctive characteristics to its progeny.

Rearing.

As it is possible to ruin the best bred cart colts by improper and insufficient keep when young, the constant personal supervision of the breeder is highly important, as a check in growth is always a loss to the owner. New laid pastures are often a hungry feed and deficient in the herbage that promotes and devolops the growth of the animal. When grazing on such land the pasturage should be supplemented by a liberal allowance of hand-feeding. An old Border sheep-farmer once remarked that "there was nothing so bad for one sheep as another," meaning of course to condemn overstocking; no greater mistake can be made than to overstock with horses; they are equally as susceptible to the evil effects as sheep, and nothing has a greater tendency to stunt their growth and development.

All foals should be thoroughly handled when young; these early lessons are seldom forgotten, and are especially useful when colts have to be castrated; besides they are much more easily and safely broken for work than when their early training has been neglected.

CHAPTER V.

THE LONDON WORK HORSE IN STREET AND STABLE.

[This chapter has been contributed by Mr. Thomas Dykes.]

THERE are few, except those who are engaged in the trade, or in the superintendence of the larger studs, who really understand the difficulty experienced in getting hold of first-class sound geldings possessing the necessary weight and strength of bone for shifting the heaviest London loads. "Were I," said the manager of one of the largest London yards, "to advertise in the Midland counties for a score of such horses, the chances are that when they were sent up on approval, I should have to consider myself lucky if I got hold of one good working pair. The others would have to be rejected as too light." To those who have visited the Shire Horse Society's Shows, since they were first established, this seems somewhat inexplicable, but if the problem were carefully worked out it would be found that were all the prize and commended stallions stationed out on the cultivable portion of the country, where mares are worked for a living when carrying a foal, one would really be astonished to find the amount of ground which has to fall under the mantle of operations of this healthy and useful movement. A result of continued agricultural depression has been evidenced in many counties in a disposition to return to what are commonly called "cheap sires," but what undoubtedly in the end must prove the dearest sires one could put to a mare. Then again there is the breeder who was thrown out by the complete falling off in the export trade, and who, having kept his colts too long uncastrated, tries to make something to pay for their keep in what little cash in service fees his neighbour can afford to give him. All that tells on the great market of London, much after the same manner in which rain-drops on a roof act in filling the water barrel. The establishments of large studs like Worsley, Elsenham, Wolferton, Dunsmore, &c., and the gathering thereinto of all the heaviest and most shapely mares, from which to breed stallions and mares for abroad or for the building up of studs elsewhere, throughout England, possibly has had a great effect in keeping down the supply of geldings at present. There exists an obvious feeling of hopefulness about this, as no doubt from these studs heavy geldings will be drawn in time to come. It must also be borne in mind that London, largely through the emulation engendered over the Cart Horse Parade movement, is cutting the old standard figures in the horse ledger, and coming up to the requirements of the times. The increased demand may make the scarcity in this way more real than apparent, but what London is prepared to pay for, and that which it will pay British farmers to supply, should not long be wanting, for granting even that the middleman, as many think, runs away with most of the profit of the business, the farmer-breeder would get a little more all the same. If the farmer should wish to breed for the London markets, then he will have to study London requirements, and when he has studied these he may be able, with a full knowledge of his land and how it must be worked at a purely agricultural profit apart from horse-raising, to come to some sound determination. The strength and richness of the pasture will have to be fully considered, for if we do get a little extra bone by using a strong, thick-legged sire, that extra thickness, which is worth £10 an inch under the knee when we bring them to market, may be lost at the mouth. The farmer on stiff, hilly clays or thin lands cannot therefore compete with those who have rich meadows, where the young colts have little to do but eat and grow big and strong. Entered for work when rising three years old, and well cared for, they develop gradually into first-class London workhorses, which, though they come in a little raw, are, in their second year of service, equal to all that the superintendent of the yard requires of them. They will on their hard feeding still continue to grow, not in height, but in width and muscular thickness, till nine years old, when their legs will begin to tell the tales of long journeys in all weathers, in a certain stiffness or grogginess and the lack of that freshness which they evidenced when first brought into the yard. Colts, very much like young cattle, if they have not been well treated when young and their growth allowed to be checked, will not so improve, however; hence the scarcity of weight in many of our street geldings in face of the heavy sires travelling may arise from what cannot be classed otherwise than as a wasteful pinching of Nature's aid and sustenance. The farmer who uses a first-class sire, and gets colts of weight which he finds (and he should find that at the first show of yearlings in his district,) he has no stock horses amongst them, should castrate early, and after that keep the gelding growing just as if he were the entire horse he hoped he would turn out. If he is not too severe on him at the outset, and gives him some good hard food, as he grows on he will be able to dispose of him readily enough when ripe to a London firm, who by continued care and attention will gradually mould a first-class horse out of him. A well-broken pair, matched as to size, colour, set or "sweep" of hocks, and regularity of step, if they have the necessary height and weight, will quite readily fetch at "five years off" £200, and if they have done useful farm work for such food as they may have eaten, heavy horse breeding ought to pay well enough. Many of them might be yoked and handled London fashion, and if so, the superintendent of the London stable could run down, try them, and take them direct off the owner's hands, thus doing away with all extra profit and expense incurred in disappointing railway journeys from fair to fair or from fair to town. Many of the best London drivers come up from the country, and though a little "green" at first, one helps the other, and once knowing the set journeys they are quite as confident as those who have driven on the stones for years. Their sons, as a rule, do not follow their fathers' occupations, the parents always looking out for something superior, as they make better wages as coopers in the breweries, millers in the large mills, or packing box makers in the manufactories, to which their fathers are attached. In regard to horses and drivers, here is a somewhat typical miller's team, driven by a very able teamsman, one who has won his diploma at the London Cart Horse Parade, also his ornamental cross—though in 1894 he was not in Regent's Park. He has to drive twelve hours a day and do all his grooming and strapping, so that his horses are under his charge in stall and stable, and it is somewhat of a treat to see how he handles them on the street. His horses are four hard browns, 16 hands 2 in., or perhaps a little over, with good blue hoofs, little hair on the leg, but well turned joints. They are rare walkers, and come round like a tandem team in the show-ring at Islington, in order that he may get up to the Metropolitan water trough. As he dismounts after the unicorn has quenched his thirst to unslip him and let up the pair, you find that he has just come up from the mills of Mr. F. D. Collen, of Bermondsey, with eighty sacks of flour, in all five and a half tons, and that the weight of waggon, loader and driver will be one and a half tons more, or a load of seven tons. This is his first journey for the day; a second with a similar load he will have in the afternoon in another direction, getting home to supper at 6.30 p.m., having left the yard on his first journey at 6.30 a.m. This from Monday to Saturday every week. The duties of the miller's horses are not of so spasmodic a character possibly, as those of the brewer's, which are, to a certain extent, affected by weather, public holidays, and the like. They are out at earlier hours than the others, as bakers are men early at work and can take delivery of sacks of flour long before the cellarmen in some of the large beer public-houses are out of bed. Of course, loads and journeys vary in the flour as well as in the beer trade, and some firms have their particular modes of harnessing and yoking. As a rule, the weight is placed next the wheels, the unicorn horse used being a light, active sort, a hundred-weight and half less than the average of the pair in rear, and worth in the market from £20 to £30 less. His powers are very severely tried at starting, but as soon as a few sacks have been delivered at the different bakeries he steps out with freedom, and if a good walker, as he ought to be, soon carries the team home for second journey at noon, or for supper and rest in the evening when work is over.

Now what should a match pair of geldings be like? That is the question the farmer should ask himself if he thinks he is in a position to raise heavy horses for the London streets. As, in the first place, they should be like each other, we shall begin with the one on the near side. He is a dark brown with black points, eight years old, 17 hands, is well seasoned, and thoroughly knows his business. We go over him as he stands without harness of any kind. His head is broad between the eyes, and his eyes have a mild, full, noble expression, suggestive of a love for his work. His chest is swelling, broad and expansive, and his short legs come to the ground with a very slight inclination inwards. The centre lines of his round, blue hoofs point straight to the front, his fetlocks are bold, firm and prominent, and proportionate to the shapely, muscular knees above. Pass round from the front, do not stand too close, and take a good view of him sideways. The head and the neck are well set on, the crest is beautifully arched, and his chin is the proper distance from where his under hame strap would fall if harnessed. The shoulders gently slope upwards to the withers, suggestive of a grand socket for the collar; the withers are not too thin, but formed so that the bottom of the collar, on which falls the strain of draught, shall be well supported at the top. The bones of the legs are flat and clad with silky feather, the pasterns possess that slope at once suggestive of support for his own body and freedom of progressive movement in front of his load. He is deep through the heart; his ribs are round as a well-hooped barrel, and the depth carried well back; his loins are broad and deeply clad with muscle, wave-like from the backbone; his quarters are broad, there is no sudden drooping, but a sabre-like sweep of outer second thighs to the hocks, which are not too wide, but suggestive of leverage without cramping cleanly chiselled out, and free from all flabbiness; his hind bone flat as in front, and his hind pasterns carried down with medium slope into the best of hoofs. Pass in rear of him and you find great, powerful inner thighs descending with mathematical evenness, all suggestive of power. Have him walked straight away from you and you notice no twisting of hock points out or in, everything being carried straight and free and parallel. As he walks back to you, you observe the same squareness of action in front. Trot him down again and he lifts his hocks cleverly every time like a bit of mechanism till you see the inside of his hoofs; bring him back and you perceive shoulder above and hoof below working as freely and evenly together. Stand to the side and see him walked. Forward he swings, five miles an hour, both ends going together, hind hoof up to old fore hoof mark, and fore-hoof launched out and on again, the pace seemingly being regulated to an inch, and never varying. Bring out his harness, and you find collar, breeching, and everything fitting like a glove. And now for the off horse. The off horse should resemble him if you can find him. It is a matter of importance in teaming, however, that the depth of the shoulders of both horses should correspond, even if they do differ an inch in height; that the style of action should be even, the paces equal in all cases, and that the hocks should be set at equal altitudes; the hind legs having a similar sweep and set. What will do for a pair will do for teams of three or four, though, as a rule, when more than two horses are used, as heavy horses as can be procured are placed next the wheels, the front pair, or the unicorn, being a little lighter in build. Still if they can be got with all the qualifications enumerated, one need never despair of finding a ready market. In regard to colour, dark brown with black points have been chosen, but bays of light or dark shades are equally suitable, and there can be no objection to good hard blacks. Greys when fully ripe, seem to be higher at the withers than others, whilst still retaining their gay carriage, and with the red roans are generally noted for their great weight. Blue roans are very rarely handsome or captivating, but on the average they have more bone than the others, and are great favourites with some London horse owners on account of their hardy constitutions and tractable dispositions. For the hard wharfinger work off the Thames on the Middlesex side, where all is sheer hard horse toil in chains and shafts, they are greatly in use. In and about the mazy wynds, and through the dark arches of Bermondsey you will come across them any hour of a hard working day, each and all walking at a faster pace than is allowed by the managers of brewery studs. On London Bridge —this article is written on the eve of the opening of the Tower Bridge, which will greatly relieve the far too congested traffic—it is interesting to watch the apparently never-ending procession passing from right to left, and left to right, each driver, from the drayman who drives his four to the costermonger on the box seat of his donkey-hauled barrow, taking his place and claiming his share of the passage. All this with good humour, though the driver who jogs the lot from the rear (and it is horses' heads to hind boards, and hind boards to heads all the way over), will come in for a good deal of strong language. The study of this moving equine democracy is at all times interesting to those who have a love for work horses. In a short time this picture like many other pictures of old London will be changed, through the lowering of the bascules of the impressing structure further down the river. Amongst the best-known work horses which pass over the bridge are those of Messrs. Samuel Taylor and Son, of Tooley Street, the well-known contractors, in whose stud of seventy strong horses there are several very grand teams of blue roans. These horses are put in at a little less price on the top standard than the horses of the larger brewing firms, but the figures run much the same on the average. Horses like the heavy massive greys of Messrs. Lewis Berger and Son, the well-known starch manufacturers, up to 17.2, with weight in proportion, and not falling away below the knee as is sometimes seen in very heavy greys, will command their own prices at any time when ripe for town work. This chapter is intended to deal chiefly with the heavier sorts of London horses, and small notice need be taken of those used in the pantechnicon vans, though these are beautiful active horses and well suited for their work, being good, steady walkers when furniture and men are all on board, and equal to trotting home with the empty van at seven miles an hour. The oil distributing people use hardy, square-legged, little cart-horses of the Norfolk type, which trot well in front of moderate loads. These horses it would pay farmers to breed on light soils, steep hill-sides, or where there was much green crop cultivation. Of what may be styled builder's horses, the heaviest, naturally enough, are those used for heavy stone hauling, and for these the greys and blue roans of Messrs. John Mowlem and Son have long been conspicuous. They must all have weight to shift weight behind. The cement, timber, and glazier and varnish trades prefer smooth-legged horses, upstanding like the Cleveland, or short and cobby, like the Norfolk cart-horses, according to the districts in which they are used. Messrs. Watney & Co. (a full notice of whose stud is given at the close of this article) possess a grand representative stud of London work horses. Messrs. Courage & Co., of the well-known Horsleydown brewery firm, have

been scarcely less noted; indeed, at the outset of the London
Cart Horse Parade, the horses of this firm, which are of the
low, square-set, blocky type, formed one of the features of the
London May Day procession. Messrs. John Watney & Co.,
of Hammersmith, have magnificient teams, which are rivalled
by those of Messrs. Young & Co., of Wandsworth, and the
fine turns-out of the Mortlake Brewery Company, some of
which are equal to the best shown in Regent's Park on Whit
Monday. Messrs. William Younger & Son, of Edinburgh,
also make a feature with their brewery horses, a good number of which are crosses of Clydesdale and Shire. Messrs.
Charrington's horses are very useful sorts; a little light
perhaps, one might think, but each and all well suited to the
particular loads and particular journeys of the firm. The
Burton-on-Trent Companies possess many fine teams, but as
a rule there is little about them to attract the attention of the
Londoner. The harnessing, equipment, and even the set of
the build of the waggon are such as to suggest reform to any
one who studies the street traffic of the Metropolis. These
large firms, however, have in many cases depôts in the
suburbs attached to railway sidings, so that the long London
journeys do not fail to be considered by the stable or stud
managers. Of the London distillery firms the most representative horses are undoubtedly those of the Thames Bank
Distillery, the leading pairs of which will average 16.3, and
this with weight and ample strength of bone. Of course,
the small family brewery horses are of the light cobby
character, not to be compared to those in the general business. Comparisons are frequently made between the work
horses of one large city or town and another; these without
regard to special conditions of labour, loads, roads, width or
narrowness of streets, or length of the journeys. Glasgow
may well be held as the city where Clydesdale work horses
are seen at their best. At any rate, the requirements of the
large Glasgow contractors to some extent rule the opinions
of the breeders, if not through the showyards, certainly

through the purchasers for the fairs and markets. Yet the English dealer for the "London stones" were in the past always prepared to go a little further at the Rutherglen, Paisley, and Glasgow fairs for a useful half dozen drawn from a string. Possibly against him a Glasgow buyer would bid for a single one, or a pair at most, and get them; but the southern men could not afford to come north and buy them by the "ones" and the "twos"; and they were no more prepared to accept light weight in the horse market-place than they would do over the grocer's counter. So the Crawfords and many others brought up the heavy Shires to breed with and produce horses of size for the southern markets. Glasgow benefited greatly, as the combination horse exactly suited the Scottish lorry, which is nothing but a Scottish, four-wheeled English waggon, such as is at present to be found in common use by Whitbread's and other firms. Messrs. Whitbread have always been partial to the "pairs" used in front of very neatly built waggons, and horses and waggons one can see very readily match. This, however, being a subject of debateable matter for controversialists on both sides of the Tweed, need not be entered into here. It is argued, however, that the present type of Clydesdale, which is largely a work of showyard and Stud Book evolution, is the best type for Glasgow work, which they say is the most severe work a draught horse can be put to. Therefore they argue this type of horse must be the best for London work also. But the journeys in Glasgow are very short; there are no "tied" publichouses, and no particular "monopolies" of the baker business amongst the millers. If we take the horses of the wellknown expert brewery firm of Wellpark, whose stud is no doubt the most representative one in Glasgow, we find that these single lorry horses carry very light loads of "stone bottled" ales, packed in barrels, to the docks, there to be shipped to India and the Colonies. In a London or Liverpool sense such horses could not be classed as *brewers'* horses; rather would they be put on the level of the horses of the carmen of the London district and suburban railway stations or the wharfingers of Bermondsey, and the south-east Surrey side of London Bridge. They have frequent short journeys out and home from the docks, but no long tiring ones over heavy roads, and do little feeding from the nose-bag.

Through the kindness of Messrs. Watney & Co., Limited, of the Stag Brewery, Pimlico, and the courtesy of Superintendent Byron, a native of the "land of cakes," hailing from the county of Ayr, but who has had extensive experience in Liverpool, the writer is enabled to give some interesting information as to their stud, which has won Shire Horse Society's premiums for four years at the London Cart Horse Parades; including leading honours for singles, pairs and unicorn teams in 1894. The full number of horses in this stud is 162; nearly all Shires of the heaviest type. The average price paid for these horses during the past ten years (1884-94) was £84. The Stud Book movement, which commenced in 1877, would seem, therefore, to have had some beneficial effect so far as the supply of geldings of the best types are concerned. Between 1880 and 1884 the Americans raised the price for entire colts which, without an export demand for breeding stock, would have found their way into the shafts. They are purchased when five and a-half years old; guaranteed sound in every way, and no horse with side-bones or ring-bones is ever selected, no matter how superior the animal may be otherwise, as the streets would soon find out the weak spot, and the exigencies of the work would not allow of their standing lame in hospital. The first three months are anxious months to the superintendent, as owing to change of climate and stable, they are frequently attacked with a form of catarrh and thickening of the glands. Their first work is generally of a light character—three half-days a week for the first three months on the shorter town journeys; but gradually, as they get accustomed to hard food, which at first consists of chopped bran and a few oats, and there is no risk of feet founder from the effects of such, they are put out on the full journeys of twelve miles out and twelve miles home, or the shorter journeys of six miles out and six home, forenoon and afternoon. The gross loads on these journeys average three tons, waggon, barrels, men, and unloading gear. In two or three years they will have put on a full hundred-weight of hard muscle, coming in at 15 cwt. 1 qr., and increasing to 16 cwt. 2 qrs., which is the present stud average. The average service for the past ten years has been six years and eight months; though there were some horses which have exceeded this by two and three years. The cost per horse for veterinary charges is 8s. 11d. per annum, and shoeing £3 10s.; a fresh set of shoes being required every three weeks. The tear and wear is more severe on the hind shoes than the fore ones, owing no doubt to the heavy friction, caused by the leverage of the hocks. The shoeing smiths meet this by using up the old "pelt" to harden the metal. The average cost to keep a horse per week, bedding included, has for the past three years been as follows:—

1891, 17s. 6¾d.; 1892, 18s. 3¼d.; 1893, 18s. 5d.

The following is their bill for diet:—

Winter Food.		Summer Food.	
Clover and Mixture	11·55	Clover and Mixture	12·25
Grains	2·80	Oats	15·00
Oats	13·00	Peas	3·15
Peas	3·15	Beans	3·15
Beans	3·15	Maize	2·25
Maize	2·25	Bran	2·10
Bran	2·10		
Total	38·00 lbs per day.	Total	37·90

Those horses which are out on the short journeys consume all their food in the stable, but those on the twelve miles journeys will have to feed from the nose-bag, and each nose-bag is filled with 20 lbs. Such horses are very liable to chills from having to stand in exposed, draughty places to disload,

SHIRE GELDINGS.

Winners of the First Prize of the London Cart Horse Parade Society; Premium from Shire Horse Society, also the Merit Badge of R.S.P.C.A., presented by the Baroness Burdett-Coutts, Whit Monday, 1894.
The Property of Messrs. Watney & Company, Limited, London.

after a hot, sharp pull; and this will possibly evidence itself at night when in stable by feverishness. A strict watch, therefore, is kept upon them, and where the temperature has greatly increased the superintendent, who must always be close to the yard, is called out at once. Naturally enough, such valuable horses are only entrusted to tried draymen, of whom there are six classes: first, 45s. per week; second, 42s.; third, 38s.; fourth, 35s.: fifth, 33s.; sixth, 32s. These draymen mostly come in from Norfolk and Essex, on the introduction of draymen friends working in the yard. The younger men have to work for six years as assistants, or in picking up "empties," before being allowed to take out a team of their own. The high character of the firm's drivers is shown every year at the London Cart Horse Parade, where they have never failed to secure the diplomas or badges of the Royal Society for the Prevention of Cruelty to Animals. In regard to the use of the heavier vans and "unicorn" teams Superintendent Dyron is somewhat against them, and the writer is with him in regard to this. The "unicorn" horse is often walking idle in chains when the "shaft horses" are doing all the work. Moreover, at street crossings where policemen give one line of traffic, turn about with the other, the extra length of horse causes delay and inconvenience to the public, and yet at the same time, owing to slack chains and their distance from the front axle, they fail to give assistance, exactly when assistance is required, to the horses behind. Pair-horse waggons, with loads to suit, Mr. Byron considers to be preferable; but the horses in these would have to be the heaviest procurable, so that they might always be equal to standing their loads. From photographs specially taken, we have pleasure in giving portrait of a pair. The brown is a Derbyshire horse of the grand "blocky" type, with great, deep, muscular shoulders; short, hard legs; well-set pasterns, and the best of hoofs. He was entered 18th November, 1889, when his weight was 15 cwt. 3 qrs.; his present weight is 16 cwt. 3 qrs., and he girths

98 inches. The other is a powerful blue roan horse of the heaviest dray type, with exceedingly strong, fine forearms, deep quarters and muscular thighs. He weighs 18 cwt. 3 qrs., and girths 96 inches. This horse has scarcely ever been a day off duty and is still quite fresh. His purchase price was £85. The farmer who can raise such geldings need never be afraid of finding a market for them in London; and at five years and a-half, they ought to yield him a handsome profit.

CHAPTER VI.

FARM MANAGEMENT OF THE HEAVY HORSE.

The various breeds of heavy draught horses have been immensely improved within a period of little more than a single decade. By the formation of breed societies and the careful and correct registration of pedigree the breeder is armed with authentic information instead of hearsay evidence and opinions often of a questionable character. Without some degree of prepotency due to inherited affinity the ancient axiom that "like produces like" frequently fails in practice. The influence of the sire generally embraces a wide field; hence the exercise of a sound judgment and discrimination is essential in selection, not only as to the leading points and general characteristics of the animal, but also as to dissecting the pedigree, and as far as practicable, tracing the merits and weaknesses of each member throughout the whole line. To the interested and intelligent breeder this information is now attainable through the stud books of the leading breeds. Up to a certain point we are in favour of close affinity; this is the best means of insuring prepotency and the fixing of a distinct type, and when skilfully conducted the system is capable of being carried out without danger of deterioration. The unsuccessful breeder is usually the man who is constantly trying the experiment of an out-cross. The stallion should be compact and evenly balanced in all his parts, standing 16 hands 2 inches on muscular, well-placed legs and sound feet. Whether for the plough or the town lorry, action is an important and valuable feature. No horse with upright shoulders and short steep pasterns can ever be a free mover. The feet is another important point. The north country adage is expressive on this point: "Feet, fetlock and feather tips may come, but bottoms never." To the intelligent breeder a careful study of the parentage will prevent many disappointments by reducing the liability to diseases of a hereditary character. The skilful breeder is careful to examine and note peculiarities of form and constitution as well as the liability of certain strains to hereditary disease. Equal attention should be exercised in the selection of the dam as regards soundness and the chances of her transmitting hereditary disease. The mare should be long, low and wide, with the limbs placed well outside the body, with free action and good temper. It is generally admitted that the difficulty of selecting the dam is equally as important as that of choosing the sire. As the physiological principles of breeding have already been discussed this subject need not again be referred to. The views of breeders have undergone considerable modification as to the age at which the mare should be put to the stud. Formerly the mare was not used for breeding purposes before the mature age of five years, but under the more liberal system of feeding which now more generally obtains she is frequently put to the stud at the age of two years, and, if well cared for, she does not suffer either in health or development. Fillies are handled and probably broken to the plough at the age of two years. Beyond this, if reserved for breeding and they prove to be in foal, they do little or no work until they are three and a half years old, when they are ready to take their share in the work of the farm. It is now generally understood that the age of all horses dates from the first day of January in each year; hence there is a growing disposition amongst breeders to contrive to have their foals dropped at a much earlier period than formerly. The practice more particularly obtains with those whose chief object is the show-

FARM MANAGEMENT OF THE HEAVY HORSE.

ring and who are generally amply provided with suitable buildings. In the case of the ordinary farmer, who is frequently placed at a disadvantage in the matter of buildings, it is best that the foal should be dropped from the first to the middle of April. When they are brought together in the show-ring, other things being equal, a discrepancy in age of three months is not easily discounted. The barren mare comes in heat early in the spring, with a periodic recurrence of the symptoms at the end of twenty-one days. It is a matter of vital importance that both sire and dam should be in the most robust health at the time of their union. The most successful result as to progeny is obtained by mating a comparatively old horse with a young mare. However much we value pedigree we prefer to have it substantiated by actual results, and we should hesitate before extensively using an untried horse. The young mare is generally more vigorous than an old animal; and the quality, if not the quantity, of the milk is much better.

MANAGEMENT OF THE IN-FOAL MARE.

The average period of gestation in the mare is eleven months, though frequently it exceeds or falls short of that period by three or four weeks. From the time of sexual connection till the time of foaling the mare may and should be regularly worked, except in the case of two-year-old mares, which, during the summer months, are grazed on a moderate store pasture. The growth of bone and muscle, rather than of fat, is the desideratum. On very poor pastures, nitrogenous foods may be used with advantage. The best winter quarters are a well sheltered grass field in which is a shelter shed or loose box enclosed by a yard. The mare should be regularly and liberally supplied with nitrogenous food and have constant access to pure water. Mares five years old and upwards may, in careful hands, be safely worked to the day of foaling. If sufficient attention is given to the supplies of food and water, nothing can be more conducive to the health of the pregnant mare. Throughout the whole period of gestation the mare should be liberally furnished with nourishing food in a concentrated form; bulky food of a low nutritive value is injurious, as it entails a debilitated system which, during the earlier period of pregnancy, frequently results in abortion.

Foaling.

When the mare has been well fed and regularly worked or exercised during the period of pregnancy the dangers attending parturition are reduced to a minimum. As the time of foaling approaches, the working mare should be placed in a roomy loose box during the night and allowed full liberty. On farms where breeding is largely carried on it will be necessary to have several foaling boxes. These are best constructed on the circular plan, which, to a certain extent, obviates the danger of accident. They should be so arranged that the attendant is enabled, not only to inspect, but to feed and water the animal without entering the box; this affords the night watcher the opportunity of inspecting without disturbing the inmate. Sometime previous to the date of foaling the food should be changed, and though still nutritive and concentrated it should be macerated with water previous to being fed. A portion of bran and linseed meal should be added; this acts as a slight and safe aperient.

The early premonitory appearance of foaling is a slight accumulation of a white adhesive substance on the ends of the teats. At this period the mare should be strictly watched both by night and day; this should be done with the greatest caution, as the mare naturally resents all interference or disturbance. Immediately the pains of labour set in a practical and careful attendant should examine and ascertain as to whether the fœtus is being presented in a natural form. If so, a moderate period should be allowed to elapse before any violent measures are used; in most cases a little well directed, though unskilled assistance may be sufficient. In every case of false presentation the services of a qualified veterinary surgeon should be obtained without a moment's delay. The cow leech and the amateur vet. should never be entrusted with the lives of valuable animals. Vigorous and prolonged convulsive efforts cannot be continued without producing injurious or fatal results in the mare. Parturition having been safely effected, the first necessary operation is to secure the umbilical cord and remove the superfluous attachment. A ligature of soft string or dressed sheep-skin should be passed tightly round near the belly and the ends securely fastened. In the case of weak foals and young mares there is sometimes difficulty in inducing the foal to suck; but a little good-natured perseverance and assistance invariably succeed.

Subsequent Management of the Mare and Foal.

It is sometimes necessary to give the youngster a small dose of castor oil to move the bowels, but when the mare has been carefully dieted this precaution is seldom required. If the dam is young the use of nutritive foods should be continued, the object being to develop the bone and muscle of the dam as well as the progeny; hence the need of food of a high albumunoid ratio, such as oats, together with a mixture of leguminous seeds and linseed. At certain periods, both the mare and foal become the victims of a serious disease, which frequently causes great mortality. Notwithstanding that the treatment should be by a skilled practitioner rather than by the practical breeder, it is within the power of the latter to ward off the attack. Septicæmia, the disease in question, is well known both to the veterinary profession and to breeders. The disease is introduced into the blood through certain organisms which abound in putrefactive solutions. Thus, for example, the foaling box may have been previously used for lambing ewes, or for a calving cow; the box not having subsequently been cleaned out becomes tainted through the decomposition of animal matter, the bacterium of putrefaction is readily communicated to the uterus of the mare by the hand of the operator during the process of foaling, or it may enter the blood of the foal through the umbilical attachment. In nine cases out of ten the disease is too far advanced to admit of successful treatment before skilled assistance is called in. The lives of hundreds of valuable animals are yearly lost by the neglect of simple sanitary arrangements, such as cleanliness, and the use of a simple and inexpensive disinfectant, such as whitewashing the walls, and occasionally sprinkling the box with dilute carbolic acid.

During the early part of the season, when the temperature is generally low, the mare and foal should be kept in the box for the first three or four days, beginning first by allowing them the range of a small well-sheltered paddock for a few hours about noon; when the weather is favourable the period of liberty may be extended. They soon become inured to the weather, whilst the outdoor exercise is conducive to health, and develops the bone and muscular powers. They should return to their night quarters until the days lengthen and the sun becomes more powerful. The artificial feeding should still be continued. At the age of a week or ten days, a slender leather head-stall is fitted up, to which is attached a short leather strap of sufficient strength to hold the young animal; this is placed on the foal and allowed to remain. A careful, good-tempered man should be told off to catch and give the foal a few short leading lessons daily; in this way its confidence is soon gained. When this is followed through the different stages of life the breaking to work is easily accomplished. The foal should also be taught at an early age to eat artificial food from the manger with its dam. Formerly, when draught horses were of less value, a few foals were bred on the large tillage farms of the Midland and Northern counties, but rarely with any degree of success. During the busy season of preparing the land and sowing the spring crops on all large tillage farms the horses are seldom in the yoke less than nine or ten hours a day. The mare comes in at noon steeped in perspiration, to be still further reduced in condition by a hungry foal. The milk is poor and unwholesome. Owing to causes such as these the foal frequently slips its hair and becomes a miserable object. The evil does not even end here. How often have we seen mares rapidly driven all day in plough or cart; when the day's work is over they probably get a hasty rub over with a straw wisp. The mare and her foal are turned into a grass field to make the best shift they can to prepare for the morrow's work. Treatment of this kind appears to be approaching near the line of cruelty to animals, of which we hear so much. Horse-breeding under such conditions as these cannot pay. In the first place, the mare suffers in health resulting in an impaired constitution and a puny progeny. The digestive and assimilative organs of a young animal can be impaired to such an extent by an insufficient supply of nutritious food as no subsequent management, however liberal, can ever restore. If breeding is to be successful the mare ought not to be worked whilst suckling her foal. One of the advantages of breeding from a two-year-old is that she is seldom put in the yoke.

Weaning.

The foal is usually weaned at the age of five or six months. Where breeding is carried on to any great extent a small, well-sheltered field should be reserved for the foals. A piece of mixed seeds containing a considerable sprinkling of cocksfoot, ryegrass and fescue will be suitable for the purpose. The grasses should be permitted to run to seed. The foals that have already learned to eat corn from the manger are delighted to have an opportunity of nibbling off the ripe seed culms. Weaning foals should never be turned out on a bare pasture, as they are liable to become affected by worms, which are difficult to eradicate, and which prevent the young animal from progressing.

As soon as the foal is weaned the mare may be gradually placed on dry food and should have one or two doses of mild aperient medicine with plenty of walking exercise or light work. For the first day or two the milk should be drawn twice daily; she should not be milked clean out, but a sufficient quantity taken to relieve her from any suffering or uneasiness. In the course of a week or less the milk will dry up.

The horse is a social animal, and pines for society. When the foal is weaned it should be placed with others of its own age. If this be not convenient, an old quiet pony, or even a donkey, makes a serviceable companion. A few loose boxes in the field, surrounded by an enclosed roomy yard, form a useful adjunct to a breeding farm. For the first day or two after weaning the foal may be confined to the yard; as soon as it gains the confidence of its companions they may be allowed to run out in the pasture during the day. We often hear the remark that foals or yearlings do not care for shelter. Be this as it may, I have always observed that those who have had the advantage of a shelter during the winter months have thriven best. The position is easily explained; as a rule, the boxes, yards, and surroundings are neglected and allowed to remain in a dirty state; while during wet weather the approaches are practically impassable. Let the yards and boxes be kept clean and dry. Let the artificial food be placed in a manger inside the box, and the foals will soon hasten there. There are from ten to twenty foals reared at Elvaston each year. They are lodged in wooden boxes with open yards enclosed by close fences constructed of old railway sleepers; they are fed early, about eight o'clock; allowed their liberty in a large grass field; and as evening draws on, as soon as they hear the voice of the attendant, they immediately race up from the farthest corner of the field.

From its earliest days the feet of the foal require unremitting attention; as the twig is bent the tree inclines; and so it is in this case. Whilst the bones are still in a cartilaginous state, through neglect of the feet, the hoofs and fetlocks, as well as the legs, often take an undesirable direction. In the case of the foal, whose hoofs are soft and tender, we prefer the use of a fine rasp to that of the draw knife; by constant attention and skilful use the feet can be well balanced—a most important and desirable point in every draught animal. Whilst the animal is still young a skilful farrier can check any tendency to irregularity. At weaning and subsequently, the leading lessons should be continued; this familiarises them to man, who soon gains their confidence.

During the early years of the young animal's life, and more particularly during the first winter, the food should be prepared. The fodder, whether hay or straw, or a mixture of the two, must be cut into fine chaff, and the corn, of whatever kind, ground into meal, the meal and chaff mixed together and well soaked with boiling water; the mass is then covered with a pliable non-conducting material and allowed to remain in this state for a period of at least twelve hours, when it will be in a suitable condition to be fed. The chief object of the breeder is to produce bone and muscle, and to do this at the least possible cost the food should be rich in albuminoids or flesh formers, rather than carbohydrates or heat and fat producers; the latter, when used to excess, impair the health and encourage the growth of flaccid muscle. The albuminoid ratio should not exceed one to four. The selection of the individual grains forming the compound will depend on their price in the market. The market value of albuminoids varies in different descriptions of grain, according as the demand is active or depressed. Sweet well-matured oats, wheat, white peas, lentils or Indian corn, and linseed should form the mixture in somewhat the following proportions: to one of oats add one-half of wheat, one-fourth of peas, one-eighth of Indian corn, and one-sixteenth of linseed. These should be mixed together in the grain and reduced to meal by being passed through an ordinary grist mill. We must not lose sight of the fact that the stomach of the horse is of limited capacity; hence it is obvious that feeding must be frequent, and to obtain the best result the food should be in a prepared and concentrated state; from four to five lbs. per day of mixed meals will be sufficient.

Another important point, second only in importance to the food, is the water. Domestic animals of all kinds, and more particularly horses, thrive best, and are more uniform in health, when they have access to running streams. This, in many cases, is impracticable on some geological formations where there are only two sources of supply. One is obtained by conserving the rain water which, under the most favourable conditions is erratic, and frequently fails for long periods. The other source of supply is obtained from deep wells; this is usually highly charged with chemical impurities derived from the rocks through which it passes. Water of this character is utterly unfitted for young horses, but by being pumped into tanks or reservoirs and exposed to the sun and air it becomes oxydised, softened and increased in temperature; and hence is better fitted to assist in promoting the animal functions. Ponds or storage reservoirs are of much value on a farm; these insure a more uniform quality. Every animal should have free access to water at all times, as when this is the case no unfavourable results are likely to follow. On this point we shall have a word to say as we proceed.

The Yearling.

As has been mentioned the age of all animals is now frequently reckoned from the first day of January in each year, although they may not have been dropped for three months later. As regards food, the same quality is continued; the only alteration is in the quantity which must be increased in order to meet the growing requirements of the animal. The leading lessons must be continued, and it is of the utmost importance that the feet should be examined and rasped down where required, at least once a month, and oftener if they show the slightest tendency to be one-sided or unshapely. Although it may be more costly we prefer a second class pasture rather than a rich feeding one, provided in the former case a fair allowance of artificial food is given. Rich grazing pastures tend to the development of fat rather than to the growth of bone and muscle.

Freedom of exercise and access to soft water, at a moderate degree of temperature, are most essential elements, conducive to robust health and progressive growth. From the middle to the end of May, the colts intended for commercial purposes should be castrated. The operation is not itself attended with much danger when it is performed by a skilful practitioner. The chief source of danger is in the casting, and this risk is now more generally recognised and the system of performing the operation without casting is more frequently practised. Many operators err in removing too much of the spermatic attachment; the testicle only should be taken; by this means more of the spirit and masculine character of the sire is retained.

In some cases docking is practised, but with the draught horse it is rather a disadvantage than otherwise. The wound, though healing over, remains tender, easily abraded, and often leads to serious accidents.

The yearling geldings and fillies are usually grazed together. The entire colts should be kept in a separate enclosure, and, if intended for exhibition purposes, they may be rather more liberally treated in the matter of artificial food; in all other respects they should be dealt with in a similar manner. Temperature and rainfall exercise a marked effect on the character and development of all animals. Low, damp soils encourage the growth of hoof, though frequently it is of a weak, spongy character. There are many popular errors as to the effect of geological formation on the growth of bone and muscle. When left to a state of nature this does, to a certain extent, obtain. The power of control is now, to a considerable extent, within the grasp of the intelligent breeder, through the selection of auxiliary feeding stuffs rich in the necessary elements of nutrition; but no selection of foods can materially alter the original framework or skeleton of the animal, the ground-work of which must be perfect.

The Two-Year-Olds.

The same treatment as that recommended for the yearlings must be continued with these, until such time as a sufficient bite of grass is obtainable, and where the land is hard-stocked or inferior in quality an allowance of artificial food should still be continued. The great danger to guard against is superfluous fat. To the eye of the inexperienced excessive obesity covers a multitude of infirmities, but with the practical man undue accumulation of fat is heavily discounted from the fact that before the animal can be brought to a healthy state, either for breeding or for ordinary serviceable purposes, the superfluous blubber must be removed. At the age of two years both fillies and geldings should be bitted and broken to the yoke.

Bitting and Breaking.

We assume that the young animals have already been trained to lead in a plain halter; then a light leather head-stall is used. This consists of the usual nose band, front piece and throat strap, one or two rings being attached to the lower part of each side. To this the bit is secured at each end by a strap and buckle. The bit should be plain, in order to prevent any chafing or injury to the mouth. By altering the side straps the position of the bit can be lowered or raised. This being accomplished the colt may be turned loose in the yard for several hours; this should be repeated for several days before any further steps are taken. At this period we prefer a round piece of hard wood of considerable circumference to the iron bit. When the colt has become sufficiently accustomed to the bit it is well to back him into a stall and have him secured on each side by a strong pillar rein. By repeating the lesson several times he becomes accustomed to and learns to be controlled by the bit. Whatever the duties of the animal may subsequently be, his usefulness largely depends on the care which has been expended on his bitting and training. An imperfectly broken horse, who has a hard uncontrollable mouth, is a dangerous brute, which, at any moment, may cause serious injury to life and limb. Having been thoroughly mouthed—and the operation should never be unduly hurried—he is next driven in reins and thoroughly accustomed to answer to the bit, to readily turn to the right hand or to the left, and accustomed to start and stop at the bidding of the breaker, while he should be further trained to answer to his name.

Having satisfactorily proceeded so far, the colt may then be harnessed in the usual way; be careful as to the fitting of the collar, otherwise you incur the danger of pinched shoulders and thus lay the foundation of jibbing. Sometimes the colt is hitched on to a tree or piece of wood, and with this incumbrance behind him is drilled up and down a grass field. When carefully bitted and broken, we prefer putting the youngster direct as body in a plough team, between two steady old horses. May I beg the reader's indulgence for a moment to explain the terms used in describing the position of the different members of a single three-horse team? The last is called the "thiller;" the middle horse the "body," and the first the "fore-horse." In most cases three horses are only doing the work of two. In a few weeks the colt becomes perfectly broken, and he is then turned out for the summer. The fillies, if well grown and intended for breeding, are stinted and turned out. Breaking and lightly working for a month or two at the age of two years improve the animal and hasten its development. At this age it is still necessary to pay constant attention to the feet. Hundreds of what would otherwise become valuable draught horses, are yearly permanently deteriorated in value by an ignorant and imperfect system of breaking. To be a successful breaker and trainer, even of draught horses, requires special qualifications of no mean order. In addition to many others, the man must be patient, firm and even-tempered. Whether actuated by the dictates of reasoning or instinctive powers the horse is not slow to resent unkind treatment bordering on cruelty.

Two-Year-Old Entire Colts.

The colt at the age of two years is capable of procreating his species, if he has been liberally treated from birth. His growth will not be impeded by serving any number of mares up to fifteen. A general characteristic in the progeny of young sires is early maturity. In the case of a draught horse this is, particularly in these times, a property not to be lightly considered. The future usefulness of the colts for breeding purposes is either made or marred; at this age they are mischievous and troublesome when turned out with other stock. One or more may be kept together in a small enclosure, with a separate box and yard for each, where they can be shut up at pleasure. At this age the colt still requires the greatest care and attention as to his feeding and exercise. Although he may have all the advantages of a roomy box and yard, voluntary exercise is of itself insufficient to maintain the colt in a healthy growing state. In addition to this, he should have at least two hours' daily walking exercise. The present show system is not conducive to the development of a healthy, well-balanced frame. Disuse does not merely relate to the lessened action of the muscles; it also entails a diminished flow of blood to the different organs of the body. The muscles can be fully developed only by constant use.

Three-Year-Old Commercial Fillies or Geldings.

The animal having now attained the age of three years, he is expected to take a share in the work of the farm. During the early spring he is taken up, fed more liberally and his former breaking lessons are revived. Where two-horse ploughing is practised he takes his place in the furrow beside a steady trained animal, and, although he may at first show some disposition to resent restraint, he eventually settles down quietly. Given a well-formed animal and the pace very much depends on his early training. A sluggish habit is more easily acquired than eradicated, either in man or animal;

hence in the draught horse the long swinging gait rather than the short quick step is to be cultivated, and this to a considerable extent is within the province of the trainer. On large tillage farms two-year-old geldings and fillies are purchased at the autumn fairs held in the breeding districts. These are wintered in the yards, and broken and prepared for the spring work. We cannot disguise the fact that few breeders who sell at one or two years take the trouble of breaking; and the work when delayed to the age of three years is seldom so satisfactorily and thoroughly performed as it is when it is systematically carried out at an early age. The purchasing and training of colts form an important part of the management of large farms; they are worked on the farm for two or three years, and are then passed on to the brewers, railway companies and town draymen. At the present moment a good sound five or six-year-old horse with power and action is worth from £70 to £100. For the best class the demand exceeds the supply. So far the production of first-rate draught horses has not suffered from foreign competition.

Feeding the Draught Horse Employed on the Land.

The efficient value of foods and the requirements of the animal are much better understood than was the case even some seven or eight years ago. We have found the following ration sufficient for ordinary farm horses, 15 h. 3 in. to 16 h. 2 in. high: 12 lbs. good sweet oats, 3½ lbs. beans or white peas, 1½ lbs. linseed, 12 lbs. short chaff, three of straw to one of hay. The corn is ground into meal, mixed with the chop, well mixed and watered to prevent waste and allowed to remain at least twelve hours in the heap before being fed; three lbs. of hay are placed in the rack the last thing at night. This we consider a full ration throughout the busy part of the year. On large tillage farms the work horses should be turned out to grass. The rotation may be so arranged that a series of catch crops may come in as required. It undoubtedly entails some extra labour, but it saves the legs of the horses, which come out much newer at the end of summer than if they had been turned out to pick up a scanty living on the bare pasture.

Hours of Work.

Except during the busy seasons, the working hours are seven in winter and nine in spring, summer and autumn, with a respite of an hour and an half at noon for rest, feeding, and watering. The limited capacity of the stomach renders the horse unfitted for long fasts without inflicting serious injury on the constitution of the animal.

Draught.

The method of attachment affects, to a considerable extent, the efficient power of the horse whether in cart or plough. In this, as in every other mechanical arrangement, the line of greatest efficiency must form a right angle to the axis of resistance. In the case of the one-horse cart the line of least resistance is maintained by elevating or lowering the length of backband, or by increasing the height of the wheels, or by raising the body of the cart by packings placed on the axle. In the case of the plough the line of draught is likewise capable of being altered by the length of the backband. The power of the horse is most efficient when not more than one-twelfth of the load rests on the animal's back. The system of tandem or yoking in line entails a large amount of waste of effective power; horses of different heights are indiscriminately yoked together; hence we have a series of different angles in the line of draught. Even were it possible to insure the full exertion of each individual, the effective power of one may be expended on the shoulder of that behind him without contributing much to moving the load. The only means of utilising the power is by direct attachment, and in the case of the plough there is no practical difficulty. By the use of equalising draught bars the resistance can be easily distributed. Another advantage presents itself in these times of dear labour, in that with three-horse ploughing a driver is no longer necessary.

Housing.

It is of the utmost importance that the stable should be roomy, well lighted, and ventilated, and free from underground drains. The building should not be less than eighteen feet wide inside; the height of the side walls should be eight feet above the level of the floor. The roof covering may either be of slates or Newcastle tiles, the latter for preference; these should be laid to a six-inch gauge, carefully and sufficiently torched inside. In cases where the covering is of slates these should be laid on $\frac{3}{4}$-inch match boards, to which the slates are secured by copper or galvanized nails. There should be a three feet passage in front of the manger for easy access in feeding and facility for keeping the manger clean. This passage should communicate with the food preparing department. The best mangers are fire clay troughs, specially prepared to pattern; a water trough of the same material is also provided. The water supply is self-acting, and so arranged that the water in the different stalls is maintained at the same level. On the same line as the manger a small hayrack is sometimes placed, though it is not always necessary.

The standings for full sized horses are six feet six inches, with a bottom and front post of oak or pitch pine, each seven inches square; the latter of which is placed in a line with the outside of the manger. Into these posts are housed and firmly secured top and bottom rails, grooved to receive $1\frac{1}{2}$-inch boards. In front of the manger, and running through the posts, are two lines of one inch gas pipe; these, when secured on each side of the posts by backnuts, make a substantial job. Before the standing posts are set and the necessary walls erected for supporting the mangers, the soil is removed to a depth of fifteen inches; the posts are then placed in position and a layer of twelve inches of hydraulic lime concrete spread over the entire surface; over this is spread a thin layer of fine concrete, consisting of Portland cement and fine granite chippings. If the work is well done the entire area becomes one solid block. A grip is formed in the concrete having connecting surface channels to receive and carry and deliver the liquid drainage on to a trapped cesspool, some distance outside the building. The only objection to this kind of flooring is its slippery character which can easily be obviated by slightly hatching the surface and radiating from the grip and carriers.

Ventilation is another important consideration. Provision should be made for the admission of fresh air on the ground line. The orifice is regulated by a slide and hence is under control; sufficient access must be provided at the apex of the roof to insure a continuous circulation. The stable should be well provided with light; this adds in no small degree to the health and comfort of the animals; but for the extra cost entailed we greatly prefer the use of boxes to that of stalls, because a hard worked animal has more freedom and rests better.

LITTERING.

Where horses are hard worked and well cared for the farmer generally takes much interest in the team. On the large tillage farms of the Border counties he makes a point of attending at suppering time, the usual hour for which is eight o'clock, and he is quite as particular as to the feeding, dressing, and watering as the stud groom is in regard to his hunters. The picker is used to the feet to clean out any soil or gravel that may have accumulated during the day. The curry-comb and brush or a hard wisp of hay or straw dipped in the water bucket, completely remove all dust from their coats

The horse-keeper next directs his efforts to the bed; the horse is moved to the near side of the standing; the litter is then carefully and regularly shaken out and a layer of clean straw is placed on the top; care is taken to arrange the straw so as to form the bed higher at the side than in the centre; the horse is then quietly moved to the off side of the standing, whilst the same operation is repeated on the near side. This simple regulation, which is strictly enforced in every well-regulated stable, enables the attendant to pass backwards and forwards to the manger without moving the animal or disturbing his bed.

From a hygienic point of view, as well as that of economy, there is no litter equal to well-cured peat moss. The quality greatly depends on the raw material, which should be that known as spagnum, or slightly decayed vegetable matter. A layer of this, four to six inches in thickness, should be spread over the floor; over this may be placed a layer of clean straw, chiefly as a preventive of dust; as a natural consequence the straw will require renewal frequently. The droppings are removed daily. The moss may be allowed to remain until it has become saturated with urine. Where this system is carefully carried out there is no liquid drainage and no decomposition to mar the comfort or endanger the health of the animals. The manure when removed can be directly applied either to grass or tillage land; the most valuable portion of the voidances are thus conserved and returned to the land, where they produce an immediate effect.

Feeding and Watering.

In comparison to the capacity of the body the horse has an extremely small stomach; hence the necessity of feeding at short intervals and fixed periods. As regards this a useful lesson may be learnt from the carman, who invariably makes long hours. At all stopping places, whether for loading or unloading he loosens the bit and hangs on the nose-bag, and although it may only be for a short time a few mouthfuls and a swallow of water are refreshing and sustaining. How often do we find the careless attendant, who is entrusted with the care of young growing animals, during the winter months when they are entirely dependent on the supplies provided for them, allowing his charges to go for months from twelve to sixteen hours from one meal to the next.

A popular error has taken hold of the public mind with regard to watering, and it is one that seems difficult to eradicate. I have never known a case of a horse injuring himself who has free access to water. It is the exhausted and fatigued animal who well knows he will be stinted, and hence drinks to excess. I have seen the practice of allowing free access carried out in some of the best managed hunting stables in England, and on extensive tillage farms where horses were kept in large numbers, and it has been attended with less illness than is usually the case. At the same time, we urge the owner of valuable horses of all kinds, to be particular as to the quality as well as the temperature of the water used. Soft water is preferable. This can rarely be obtained, except through the agency of running streams or where water is impounded in reservoirs, and is airified and softened by the action of the wind and atmospheric air. If possible, the temperature should not be below 55 degrees.

Grooming.

This is an important department of stable management which is far too often neglected in the case of the draught horse. Next to feeding and watering it is the point that is most conducive to health and comfort. Good grooming promotes healthy circulation and tends to carry off noxious humours from the body. How often do we see horses coming out to work in the morning with coats staring and full of dust. Pass the hand down the chest and between the fore-legs, and you find the hair matted with the perspiration of previous days or weeks. The draught horse should be as carefully dressed twice every day as the conditioned hunter. In my boyish days the Clydesdale farm horses on Carrick Shore were a sight that would well repay a long journey. To see the men in their white mole-skin trousers and waistcoats; their horses as sleek and fine in their coats as the best carriage horses, with every chain and buckle shining like silver, afforded great pleasure. Since then the horses are much improved, whilst I regret to confess that they are not turned out in the same state as formerly. Here the labour difficulty is as keenly felt as it is in the South.

Shoeing.

This belongs to the province of the veterinary surgeon rather than that of the practical agriculturist. I cannot, however, resist the temptation of a word on so important a subject. There is an old adage, "No foot, no horse." The writer much prefers flat shoes to calkins. The shoe should, as far as practicable, be fitted to the foot, not the foot to the shoe; keep the heels low, as this promotes expansion. Under no circumstances should either the bars or sole be cut away with the knife; the walls or horny crust of the hoof should be levelled by the use of the knife or rasp, in order to form a level seating. The shoe may be slightly burnt in order to allow for the natural expansion of the hoof. The heel of the shoe, for a space of two inches on each side, should be slightly bent away from the hoof; the walls of the hoof should take a bearing of about three-eighths of an inch all round on the shoe. Most shoeing-smiths are apt to use too many nails, which has a tendency to weaken and injure; they generally use three nails on the inside and four on the outside. We think two good nails on the inside and three on the outside would generally be sufficient. The shoes should not be fullered, but the nail-holes countersunk in the solid; under no pretence whatever should the rasp be used on the outside crust of hoof, as this is easily injured. When the outside integuments are broken moisture is allowed to enter, and weakens and rots the hoof.

Feeding Town Horses.

The horses belonging to town draymen and which are used chiefly for the removal of heavy merchandise, are driven quicker and are generally heavier loaded than those belonging to the large London brewers. Those who are interested in heavy draught horses are surprised that so great a difference exists between the horses belonging to the large railway companies as well as those in use by the Glasgow draymen and those of the large London brewers. Amongst those who are engaged in the commercial enterprise of the large provincial centres of industry, the Northern horses, though not more weighty, are regarded as more muscular, and are carefully selected for the development of those points which insure free action. Take Glasgow as an illustration of the Northern division and compare it with London as the chief centre of the South. In the former case the average weight of the dray horse is 16 to 18 cwt., whilst in London the best dray horses vary from 18 to 21 cwt. The hours of labour are practically the same, though the loads are widely different. The average load of a single horse on the streets of Glasgow is, inclusive of the dray, three tons, ten cwt., whilst in London it is a common occurrence to see two ponderous horses tugging at a load of five tons. Probably the system of yoking double has much to answer for. Some of the best work horses are to be found in Liverpool. In this connection it is only fair to say that the average period of the life of a dray horse is greater in London than in Glasgow; this we attribute mainly to the different systems of feeding, to which we will shortly refer. The usual hours of work of town horses is not less than ten, and this in all weathers. In all well managed stables horse-keepers are employed whose business it is to clean, feed, and harness a fixed number of horses, so that on the strike of the

ENGLISH DRAY HORSES.
From a Painting by J. C. Zouter (1810).
In Sir Walter Gilbey's Collection.

clock they start from the stable. The horsekeepers must be regular in their habits and early risers. In large stables they are under the close supervision of a foreman. The stablemen's time is occupied during the day in thoroughly cleaning the stables and mixing and preparing the food for the mid-day, evening and morning meals. For a dray horse in full work the average weight of dry food per day is 36 lbs. The materials of which this ration is composed are subject to wide variation, sometimes without due consideration as to the nutritive value of the different varieties and the purposes they have to serve in the animal economy. We have noted the effects on hard worked tram horses of a ration composed of 10 lbs. Indian corn, 2 lbs. beans, 4 lbs. oats and 6 lbs. long hay per day. It is needless to say that under this treatment the company soon required to recruit their stud. In some stables the daily ration is: oats, 8 lbs., Indian corn, 5 lbs., beans, 4 lbs., bran, 4 lbs., with 16 lbs. hay chaff: in other stables barley is partly substituted for Indian corn. In the Glasgow stables boiled turnips are used to some extent; to this cause we attribute the great mortality amongst these horses. The writer can, from experience, recommend the following as a maximum daily ration for a dray horse in full work :—16 lbs. oats, 4 lbs. white peas, 14 lbs. hay-chaff cut fine, carefully sifted to insure freedom from dust, 2 lbs. linseed, 4 lbs. long hay. It is essential that the grain be reduced to meal, then mix with the chaff, liberally sprinkle the mass with boiling water, turn over several times in order to insure the saturation of the dry meal and cause it to adhere to the chaff; allow the mixture to remain at least twelve hours before being fed. One of the most troublesome ailments to which horses are liable when fed exclusively on dry food is indigestion or colic. By adding a small quantity of linseed and slightly macerating the food, this difficulty may be entirely obviated.

A detailed account of the best systems of management of London work horses is given in the preceding chapter.

Harnessing and Yoking.

The harness is of the utmost importance. It should be of the best materials, strong yet light. Every part of it should correctly fit the animal; here again we have a widely different practice between that which obtains in the North and that of the South; in the former a light head-stall is used, strong but light; this familiarises the animal with the different objects as they come within his range of vision, hence the liability of accident is less common. Again the collar should be a true fit, and should be carefully attended to from time to time. The top is light, and well shaped, either with or without ornament, the haimes should suitably embrace the collar; the hook, or tug attachment should be so constructed that by a simple arrangement the line of draught admits of being altered several inches. Another important point is the tree of the cart saddle. The upper curve for carrying the back-band should be lined throughout with a thin plate of steel; this when kept well lubricated is not only more comfortable, but it increases the efficiency of the horse. As with many another practice, fashion rather than utility rules. The heavy bit halters of the South with their bearing reins; the heavy collars surmounted by ponderous housing like the mail armour of the ancient warrior are of themselves no insignificant load. The curved groove of the saddle tree is armed with probably only three clips which do not afford free play for the back-band. The angle of draught is also a point which does not receive the amount of attention which its importance imperatively demands.

CHAPTER VII.

DISEASES AND INJURIES TO WHICH HEAVY HORSES ARE LIABLE.

IN the following remarks on the diseases and injuries to which the heavier breeds of horses are liable in a state of domestication, it is not intended to give such information as will enable the horse-owner to play the part of veterinary surgeon, and treat his animals in every case as if he were a person thoroughly trained in veterinary medicine and surgery. Such endeavour would be as futile as it would be inexpedient and dangerous. Printed directions and horse-doctor books cannot do this; the utmost service they can yield is to afford the attendant upon, or the owner of horses some idea of the disorders and accidents to which these creatures are exposed, so that he may be able to form an idea as to what should be done before the arrival of the veterinary surgeon, in cases of emergency, or when the assistance of this useful individual cannot be readily obtained. The majority of horsemen now-a-days have received some kind of instruction in horse-management, either at one of the several agricultural schools established in the United Kingdom, or by attending the lectures and demonstrations so frequently given in various parts of the country, through the instrumentality of agricultural societies or County Councils. In any case, for the treatment of the more serious diseases and accidents far more experience and skill are needed than are possessed by the amateur, however well read he may be in veterinary books; so that, in order to avert loss and damage, it is the wisest course to invoke professional aid without delay, resorting to such measures as may be deemed appropriate until its arrival.

For this reason only a few of the more frequent diseases and accidents will be referred to, and these briefly.

Diseases.

Fever.

Symptoms.—Fever is a condition of the body in which the temperature is higher than in health. The ordinary temperature of the horse's body—what is termed the internal temperature—is about 100° Fahrenheit. It is best ascertained by the self-registering thermometer, which is inserted into the rectum and kept there for a minute or so. When this temperature rises above 101° fever is present; if it reaches 104° then the fever is somewhat serious, and when it gets to 106° it is very severe. In proportion to its height the horse becomes wasted and debilitated.

The pulse—which is usually 38 or 40 beats a minute, and is best felt inside the lower jaw—is correspondingly increased, and the beats may reach 60, 80, or even 100 per minute, though when it is over 80, the fever may be said to be high. The breathing is also quickened, the number of respirations—which are about 8 per minute in health—increasing in a corresponding manner with the pulse. Coincidently with these phenomena the skin is dry and hot, though exceptionally it may be wet with perspiration. The mouth is also dry, hot and pastey when the finger is passed into it, and it generally has the odour of indigestion. The appetite is either much diminished or lost; and though the horse may drink a good deal of water, the urine may be less in quantity and high coloured. Sometimes the breath feels very warm, and the eyelids are swollen, with perhaps tears running down the face; in certain cases the horse is somewhat excited, in others he is listless, apathetic and depressed.

Fevers are of several kinds—such as continuous, remittent and intermittent, according to their course; and simple, specific, inflammatory, adynamic or hectic, according to its symptoms and cause.

In many cases, at the very commencement of fever there are signs of rigor or chill, the coat being then lustreless and hair erect, and the skin cold, wholly or in parts; while the horse may even be trembling slightly. The diminished desire for, or refusal of food is always a very significant sign of commencing illness in a horse, and should therefore receive immediate attention.

Treatment.—The causes of fever are numerous, and its successful treatment largely depends upon the cause being ascertained. This is discovered by noting the symptoms and inquiring into the history of the case. This needs tact and skill, and as some of the fevers are very serious and soon run on to a fatal termination, it is advisable to obtain veterinary advice in good time. The amateur, however, can assist in the treatment by having the horse moved into a well ventilated loose-box or stable and made comfortable, but not oppressed, by means of clothing and bandages to the legs; if the latter and the ears are cold, which is sometimes the case, then they should be hand-rubbed. The horse ought to be allowed plenty of cold or tepid water to drink, with sloppy food. Nursing is the chief means by which restoration to health can be secured. Medicines must be sparingly given by the unskilled, and at most nothing more should be administered than about an ounce of nitrate or carbonate of potass in a bucket of water once or twice a day. If he will lie down, the horse should have a good soft bed. He ought not to be exercised until the appetite has returned, nor put to work until he feeds well, and has regained his usual strength and spirits.

As nearly all young horses brought up from grass or from

the country to town stables are liable to an attack of town or stable fever, they should be put into airy stables, and carefully fed and exercised until they have become somewhat seasoned. And when put to work this should be light and only for a short time at first.

Catarrh.

Catarrh, or what is termed a "cold in the head," may attack old and young horses alike, and at any season of the year; though it is most frequent in cold or changeable weather. One of the great predisposing causes is a hot and badly ventilated stable.

Symptoms.—There is more or less fever at first, with sneezing, perhaps shivering, cold legs and listlessness, and slight loss of appetite. Soon there is a discharge of watery fluid from the nostrils, sometimes also from the eyes; this later becomes yellow and purulent, and not unfrequently cough ensues, with sore throat and more or less difficulty in swallowing. Very often, too, these symptoms are accompanied by more fever, loss of appetite, and swollen glands about the upper part of the throat.

Treatment.—The treatment chiefly lies in nursing, making the horse comfortable by body clothing and leg bandages, keeping the stable at a moderate temperature and well ventilated, and giving mashes of bran and linseed, with small quantities of nitrate of potass in the drinking water. The head may be held over a bucket of boiling water in which there is some hay and a little oil of turpentine or carbolic acid, so that the steam may pass up into the nostrils. If the cough is troublesome, the upper part of the throat may be well rubbed with soap liniment, or a liniment composed of equal parts of olive oil, oil of turpentine and spirit of hartshorn. Should the cough be very severe, a little tincture of opium or chloroform may be dropped into the bucket of hot water, and a sack or blanket thrown over it and the horse's head, in order to keep in the vapour.

Should the horse be debilitated after the more severe symptoms have disappeared, a dram of powdered sulphate of iron may be mixed in the mash once a day.

Strangles.

This is an infectious disease to which young horses are more especially predisposed, and somewhat resembles catarrh. One attack generally protects horses against a second. There is great probability that every case of strangles is due to infection, and from this point of view, and also because of the trouble and damage it only too often occasions, it should be treated as a communicable disease, so as to prevent its spreading.

Symptoms.—It generally commences with fever and dulness, and disinclination to eat. The throat begins to feel sore and there is difficulty in swallowing; while the glands between the jaws and below the ear are swollen and painful to the touch. In nearly all cases there is inflammation of the air-passages of the head, and this is manifested by a discharge of yellowish matter from the nostrils; there may also be cough. The swelling between the jaws increases in extent and painfulness, and not unfrequently this causes obstruction to the breathing, which is marked by a noise both in inspiration and expiration. In some cases this obstruction is so great that suffocation is imminent, and to prevent it the wind-pipe has to be opened lower down the neck and a tube inserted through which the horse can breathe. In the usual course of the disease an abscess forms in the middle of the swelling, and when this bursts the horse is generally relieved; the swelling subsides, fever rapidly diminishes, swallowing becomes easier and the appetite is increased.

This is the ordinary course of the disease, but sometimes it runs an irregular course. The fever persists, and the other symptoms may increase in intensity; swellings appear in different parts, and these may form abscesses, or disappear and again reappear elsewhere, and the disease may continue for a very long time—in the simple form it seldom lasts longer than a fortnight or three weeks, whereas in this malignant or irregular form it may run on for one or two months, or even longer. In some instances it is a year or two before the animal completely regains a healthy and robust condition. This protracted phase of the disease is due to the repeated occurrence of abscesses in various parts of the body—these suppurate, heal up, and are succeeded by others; they sometimes form in the internal organs and then usually cause death.

A not unfrequent sequel of strangles is "roaring," which greatly depreciates the animal's value, as it interferes with the breathing.

Prevention.—Strangles should be dealt with as a very contagious disease, and careful isolation of those affected, with disinfection measures, ought to be strictly observed.

Treatment.—Good nursing must form the chief part of the treatment of strangles. Whenever a young horse shows signs of ailing, it should be placed in a well-ventilated and moderately warm stable or loose box—the latter is always the better; this should be kept clean and comfortable. If the weather is chilly a blanket may be worn over the body, and it may even be necessary to place woollen bandages on the legs if they have a tendency to become cold. The food should be soft, and consist of bran and linseed mashes, oatmeal gruel, and a little good meadow hay, with now and again some scalded oats. If in season, grass and carrots, or sliced turnips, are good. The water given to drink should have the chill taken off if the weather be cold, or the oatmeal gruel may suffice. A little nitrate of potass—say half an ounce—may be put in the drink now and again.

If the fever runs high, a fever draught may be given (this is also desirable in ordinary fever and catarrh); a useful draught is composed of acetate of ammonium, in solution, three or four fluid ounces; sweet spirits of nitre, one ounce; bicarbonate of potass, half an ounce, to be mixed in a pint of tepid water. This draught may be given once a day until the fever abates. Should the breathing become noisy, or the horse experience much difficulty in swallowing, then hot water vapour should be inhaled as in the treatment of catarrh, a little carbolic acid or oil of turpentine being added to the water. At the same time, the white liniment recommended for sore throat in catarrh should be applied to the upper part of the throat and beneath the jaws where the swelling takes place. Sometimes, when the swelling is very extensive and dense, it is well to apply a hot linseed meal and bran poultice to it, or to blister it with cantharides ointment.

The abscess may be opened when it is fully formed, which is ascertained by its "pointing" and feeling very soft at a certain part, or left to open spontaneously, which is the best unless the amateur is sufficiently skilled in using a lancet. When it is opened, the wound should be kept very clean by washing with warm water and a sponge, and dressing with a solution of carbolic acid—one to fifty of water.

If the fever has been high or the abscess large, there is often a good deal of debility supervening, and this must be combated by a generous diet, such as scalded oats and boiled linseed, to which some salt has been added. If there is very much prostration and the digestion is impaired, it may be necessary to give a pint of milk two or three times a day; to this a tea-spoonful of carbonate of soda should be added. Sometimes it has been found advantageous to give one or two eggs beaten up in milk in the course of the day, or a pint or quart of stout or porter morning and evening. In the irregular form of strangles the same system of nursing should be carried out, and the abscesses opened wherever they appear. Sulphite or salicylate of sodium may be given in half ounce doses in water twice a day. A stable or loose box which has been occupied by a horse affected with strangles should not be again used until it has been thoroughly cleansed and disinfected.

Influenza.

This is undoubtedly an infectious fever, which appears in a very extensive manner over large tracts of country, the outbreaks always occurring where there is much movement of horses from one place to another. In this way it follows the lines of traffic, and may appear at any season of the year; horses of all ages and under all kinds of conditions, may be affected, but it generally visits most severely those which are badly attended to and kept in unhealthy stables.

Symptoms.—The most marked characteristic of influenza is the intense prostration that accompanies the fever from the very commencement. Otherwise, in most of the outbreaks the symptoms are much the same as those of catarrh, and they may all be developed very quickly. Sometimes the air passages and lungs are chiefly implicated; at other times the abdominal organs suffer most, and in some of the outbreaks symptoms of rheumatism, with swelling of the legs, head, and other parts of the body, predominate. Not unfrequently we may have all these symptoms manifested by one animal. The disease has received several names according to the prevailing symptoms. The catarrhal symptoms may be well marked, and then we have, in addition to the fever and great debility, the signs of ordinary catarrh; these, under favourable conditions, gradually subside in eight or ten days, and in a fortnight or three weeks the animal has usually recovered.

When the lungs and bowels are implicated, however, the cases are more serious, especially if the sanitary conditions are bad and the horses are not healthy and vigorous.

Treatment.—One of the essential conditions in the successful treatment of influenza, is relieving the animal from fatigue and work whenever the first signs of illness become apparent. These signs are generally diminished appetite, listlessness, weakness, dry hot mouth, hanging head, swollen eyes, and perhaps shivering. To work and fatigue the horse after the disease has seized him, is to expose him to the risk of a more severe attack than he otherwise would have, and may lead to his death.

Therefore cessation of work at once is all important. Good nursing comes next in importance, for the amateur—and even the veterinary surgeon, for that matter—can do little more than place the patient in the best possible hygienic conditions and maintain the strength. More horses are injured than benefited by the injudicious administration of drugs in this and many other diseases.

Good ventilation, keeping the horse's body warm and comfortable, and giving soft and easily digested food, are the chief points to be attended to. If the symptoms are mainly those of catarrh, then the treatment should be the same; if the chest is affected, then the treatment should be the same as for pleurisy or inflammation of the lungs; and when the bowels are implicated the treatment prescribed for inflammation of them must be adopted. When the legs and other parts of the body swell, then they should be kept as warm as possible by means of woollen bandages and rugs. Salicylic acid should be given in one-drachm doses in a little thick gruel twice a day. When the animal is recovering, in order to counteract the debility, it is advisable to give vegetable and mineral tonics. The best of those for the horse are powdered gentian and sulphate of iron—an ounce of the first and two drachms of the second—in ball, once or twice a day. Boiled linseed is advantageous.

The horse should not be put to work until quite recovered, and even then this should be rather light for some time.

Glanders and Farcy.

These are not two diseases, but only one disease in two forms. We shall, therefore, treat of these as one disorder under the name of glanders.

Glanders is a virulent disease special to horses and asses, but transferable from them to several other species of animals, and to mankind. It may affect every part of the body, but is most frequently witnessed in the head, and on the skin. It may be chronic or acute, but it is generally the former, though both are marked by fever, which is most severe in acute glanders. It is very contagious, and can be produced by giving the poison in the food, or water, or in a ball, and it can gain introduction to the system by inoculation, through a wound or abrasion, and in other ways. Contact with glandered horses, being put into stables which have been inhabited by them, drinking out of water troughs they may have frequented, or eating from receptacles they have fed in, are the usual ways in which healthy horses acquire the disease. It is most frequently witnessed among large studs of horses, and especially those which are overworked, improperly fed, or badly housed. Low condition predisposes to, but cannot generate the disease. A variable period elapses between an animal's receiving the poison and the appearance of the first symptoms, but it is between a week and several months. The poison is contained in the discharge from the nostrils, and in that from the sores, as well as in the blood and other fluids; but the disease is mainly spread by means of the matter from the nostrils and sores. In the ass and mule glanders nearly always appears in the acute form, and rapidly runs its course.

Symptoms.—The symptoms in acute glanders are much more marked than in the chronic form, but the high fever constitutes the chief difference. This fever lasts for a few days generally, then subsides, but only to reappear after a short interval. There is much depression, and the animal does not care to move. There is usually a discharge of a yellowish sticky matter from one or both nostrils, which adheres around them, and at the same time there is one or more sores, or ulcers inside the nostril on the partition separating the nostrils. If the discharge is only from one nostril, then the sores are on that side. When the ulcers are deep, then the discharge may be streaked with blood. The glands inside the lower jaw are also enlarged, hard and knotty. Ulcers may or may not appear on the skin at the same time. Sometimes the ulcers are high up in the nostril, and cannot be seen, and not unfrequently they extend down the windpipe. The lungs are generally implicated, or they may be above the seat of disease, but this is more frequently the case in chronic glanders. In the acute form, if the horse is not killed it dies from suffocation or exhaustion.

The chronic form only differs from the acute by the severity of the symptoms. A horse may live for a considerable time when affected with chronic glanders, and even perform hard work, as the constitutional symptoms are comparatively slight. But the disease always terminates in acute glanders if the horse is not destroyed.

Farcy is merely superficial or skin glanders, and it also may be acute or chronic. There are ulcers on various parts of the body, and these generally discharge; they are connected by a prominent line or "cord." The legs are most frequently involved, and then they are generally swollen and painful, and the horse moves with difficulty. Farcy generally terminates in glanders.

Treatment.—Glanders is practically incurable, and owing to its dangerous character its cure should not be attempted. Diseased horses should be at once destroyed, and those with which they have been in contact, or which have stood in the same stable with them, ought to be considered suspected, and consequently kept apart from others. Stalls and places which have been occupied by diseased and suspected horses should be thoroughly cleansed and disinfected.

Bronchitis.

Bronchitis is inflammation of the lining of the windpipe and its branches in the lungs, and is usually due to colds, though it is sometimes a complication of other diseases, and it may even be produced by the entrance into the air-passages of irritant fluids or gases.

Symptoms.—Bronchitis may be acute or chronic, but in young horses it is most frequently the former. This generally begins with shivering and dulness; then fever sets in, and the breathing is quickened, while there is a hard, loud and frequent cough. There may or may not be a discharge from the nostrils at first, but there is generally after a day or two, and in a few days it may be quite copious. The cough increases in frequency and severity, and is very exhausting, while the appetite is much diminished. Death may ensue from filling up of the bronchial tubes with matter. But a favourable result may be anticipated when the fever gradually subsides, the cough becomes softer and less frequent, and the discharge from the nostrils less and thinner in consistency.

Chronic bronchitis is generally seen in old horses. There is little, if any fever, and the nasal discharge is very trifling, the most marked symptom being the cough, which is often very harassing.

Treatment.—As bronchitis commonly occurs in cold weather, the horse should, if possible, be put into a comfortable, well-ventilated stable or loose box, and the body clothed, the legs being enveloped in woollen bandages or straw or hay bands, after being well hand-rubbed. Hot water vapour, into which a small quantity of oil of turpentine or carbolic acid should be put, ought to be inhaled by the animal, as for catarrh; and the throat should be rubbed with the white liniment already mentioned, or with compound camphor liniment. The same liniment may also be applied to the sides of the chest, or this may be enveloped in a thick blanket and hot water (not scalding) poured on it for an hour or two at a time; the blanket must then be removed, the skin thoroughly dried, the liniment rubbed in, and a dry blanket put on.

A draught composed of one drachm of camphor, two ounces of solution of acetate of ammonium and an ounce of nitric ether, mixed up in about ten ounces of water, should be administered twice or three times a day. The diet should consist of mashes of linseed and bran, with a few scalded oats;

carrots or green food should also be allowed, and a little good hay. When convalescence is setting in a drachm of powdered sulphate of iron may be given in the mash twice a day, and the food may be more nutritious.

Little can be done for chronic bronchitis beyond keeping the horse in a cool, well-ventilated stable, clothing the body comfortably, giving easily digested food, and allowing steady slow work.

Congestion of the Lungs.

No animal is so liable to congestion of the lungs as the horse, and it may be an accompaniment or sequel of other diseases, or occur by itself. It usually appears in the acute form in the latter case, and it is this which will now be noticed.

Acute congestion of the lungs may be induced by sudden severe exertion when an animal is not in good condition, or by long-continued severe exertion even when in good training; it may also be caused by exposure to cold, and especially to cold winds and wet.

Symptoms.—The symptoms of acute congestion of the lungs are of a very decided character. The breathing is extremely hurried and laboured, the nostrils are widely dilated, head carried low, countenance anxious and haggard, body usually covered with perspiration, legs stretched out and cold, the flanks heaving tumultuously, and sometimes the heart can be heard beating violently. Not unfrequently blood flows from the nostrils, and if this is foamy it shows that it comes from the lungs. If not quickly relieved the horse will die from suffocation.

Treatment.—This, to be effective, must be prompt. The horse should not be moved or disturbed, and if wearing harness this ought to be taken off. An abundance of fresh air must be allowed; the legs and body should be well rubbed and clothed, and if any turpentine liniment is at hand it should be applied to the legs before they are bandaged.

Brandy or whisky, in six-ounce doses, may be given in water every hour or two hours for the first three doses, and then every four hours for four or five doses. If there is thirst, cold water, or, better, oatmeal gruel can be given. If the symptoms do not soon subside, hot water should be applied to the sides in the manner already indicated, and care should be taken to keep the animal from draughts of air.

After recovery some days' rest should be allowed, and careful feeding observed.

Inflammation of the Lungs.

Inflammation of the lungs may be a disease of itself, or it may follow catarrh, bronchitis, congestion of the lungs, or other disorder, as well as be due to sudden chill, foul hot air in stables, &c. Pleurisy is often present.

Symptoms.—There is fever, the pulse and respirations are increased, the animal is dull and dejected, and wanders about in the loose box, but rarely lies down. There is frequently a short, dry cough, and there may also be a slight discharge from the nostrils of rust-coloured mucus when the disease is advanced; the skin of the body and the legs are cold, the mouth is hot and dry, and the membrane lining the eyelids and nostrils is deep red in colour.

Treatment.—This is similar to that for congestion of the lungs. Fresh air is above all things necessary; at the same time the body and legs must be kept warm. From four to six quarts of blood abstracted from the jugular vein sometimes lead to a favourable change in the case of fat, high-conditioned horses. At first the following draught may be given every four hours:—Fleming's tincture of aconite, six minims; nitric ether, one ounce; solution of acetate of ammonia, four ounces. To be given in a quart of thin gruel or tepid water.

If there is much debility, then instead of this draught, six ounces of brandy or whisky may be administered three or four times a day in the same manner.

The food should be sloppy mashes of bran or linseed, with oatmeal gruel, a little good hay, green forage, or carrots. Cold or tepid water may be allowed to drink, and in a bucketful of it an ounce of nitre may be dissolved.

When the horse is recovering, a drachm of powdered sulphate of iron may be given in the mash twice a day.

Pleurisy.

This is inflammation of the membrane lining the chest and covering the lungs, and may be a complication of pneumonia or other diseases, or exist independently.

Symptoms.—There is fever succeeding a shivering fit. There is most acute pain on moving the ribs, which causes the horse to keep them fixed as much as possible, and to breathe quickly, in a careful manner, with the abdominal muscles. The countenance looks distressed, and there is a short interrupted cough, while on attempting to turn there is heard a painful grunt. Pressure between the ribs causes acute pain; the horse does not lie down. Effusion into the chest very often sets in early, and then there is less pain, but the breathing becomes deeper and laboured, owing to the pressure on the lungs.

Treatment.—This does not differ much from that adopted in inflammation of the lungs. The general management should be the same, and the hot water applications to the chest should be even longer continued. Mustard may be applied to the sides of the chest with advantage. Nitrate of potass, in ounce doses, should be given in the water or gruel, and Fleming's tincture of aconite, in four to six-minim doses given in a small quantity of water every three or four hours.

After three or four days, whisky in four-ounce doses may be given twice or three times a day in gruel. If fluid accumulates in the chest, then it should be gradually removed by surgical operation, which the amateur had better not undertake.

Rheumatism.

Some horses are particularly liable to rheumatism, which is an inflammatory condition of certain structures in connection with joints, tendons, muscles, &c.

Symptoms.—Rheumatism may be acute or chronic; the acute form is accompanied with fever, and usually manifests itself suddenly in the joints of the limbs—as the stifle, fetlock, hock, knee, or sheaths of the tendons. There is great lameness and pain on pressure, and often more than one part is affected; not unfrequently the swelling and pain leave the joint as suddenly as they came, and attack another part. The heart is often involved. In bad cases the joints are much damaged.

Treatment.—Hot fomentations to the inflamed parts, of water in which poppy heads have been steeped; with gruel, in which ounce doses of the bicarbonate of potass have been dissolved. The animal should be kept comfortable, and if there is constipation, a mild dose of physic may be given. If the fever runs high, salicylate of sodium in two-drachm doses, three times a day, should be given in a pint of water or gruel. When the inflammation in the joints or sheaths of the tendons becomes chronic, then it may be necessary to rub them with the white or soap liniment, or with the following liniment:— Coutts' acetic acid, two ounces; whisky, two ounces; oil of turpentine, two ounces. One white of egg to be beaten up with these. The skin should be first well brushed, then the liniment firmly rubbed in.

Laminitis.

Heavy horses are more liable to inflammation of the feet, perhaps, than light ones; and the fore-feet are much oftener affected than the hind ones. Many causes will give rise to it, such as bad shoeing, injuries, severe travelling in hot weather, indigestion, superpurgation, &c., while it is often a sequel of pneumonia, influenza, &c.

Symptoms.—This is a most painful disease, and is accompanied by a considerable amount of fever. The horse perspires, breathes quickly and looks as if suffering agony; the symptoms might be mistaken for inflammation of the lungs but attempting to make the horse move reveals the nature of the disease. He will not stir if he can avoid it, but remains rooted to the ground, resting his weight as much as possible on the heels. The feet feel extremely hot, and tapping the hoofs intensifies the pain.

Treatment.—The shoes should be removed from the inflamed feet, if possible, and the walls lowered to a level with the soles, so as to allow these and the frogs to sustain a greater portion of the weight. But this is a difficult operation, as the horse suffers excruciating pain when one fore-foot is lifted. The animal should therefore be put into a sling, or better, thrown down, the litter being peat moss or sawdust. This allows the shoes to be taken off and the feet to be attended to. Cold poultices of bran or other material, or cold wet clothes, should be applied to these and kept constantly wet and cold. Carbonate of soda may be mixed with the poultices or water. Unless there has been purging, a dose of physic should be given, and the diet ought to be of a laxative nature. If the horse is lying and does not attempt to change position, he should be turned over every day to prevent the occurrence of sores on salient parts of the body. When the intense pain and inflammation have subsided, exercise on soft ground should be enforced for some time.

Colic.

Colic is spasm of the intestines, or it may be due to distension of these with gas (flatulent colic). Many causes may give rise to colic—such as indigestion, mismanagement in feeding or watering, chills, worms, &c.

Symptoms.—The attack is usually sudden, and the chief sign is the manifestation of restlessness, owing to the pain experienced. The horse lies down and rolls about, then gets up, shakes himself, looks towards his flanks, paws, strikes at his belly with the hind feet, and if in a loose box wanders around it. The pain subsides, and the horse then remains quiet and may commence to eat, but in a short time the symptoms reappear, and at each recurrence they may increase in intensity; attempts may be made to stale, while the animal generally perspires freely and manifests anxiety. In flatulent colic the symptoms are analogous to those in spasmodic colic, the chief difference being that in the former there is distention of the belly, and the breathing is therefore more interfered with; the horse also lies down more carefully and does not roll so much.

Treatment.—No time should be lost in treating cases of colic, and the relief of pain is one of the first objects to be obtained. Six ounces of whisky should be given in a quart of tepid water, and if two ounces of laudanum can be added to this, so much the better. The belly should also be well rubbed with straw wisps. If there is constipation, a dose of physic ought to be given; and when there is distention of the abdomen, after the stimulant just mentioned an ounce of oil of turpentine in a pint of linseed oil ought to be administered. The alcohol and laudanum may be repeated in three or four hours if the symptoms do not abate. An enema of soap and water every two hours is very serviceable in obstinate cases; and when the attack is acute, blankets wrung out of very hot water and applied to the abdomen often act very beneficially.

Inflammation of the Bowels.

Like colic, which it often succeeds, inflammation of the bowels arises from many causes.

Symptoms.—These are not unlike those of colic, except that there is no intermission in the pain, which is much more severe, and the breathing and pulse are quickened throughout; the pain is also increased by pressure on the abdomen. The body is covered with profuse perspiration, and the expression is haggard and distressed.

In this disease no alcohol should be given, nor yet laudanum; but, instead, powdered opium in two or three-drachm doses, rubbed up in flour gruel, every two or three hours; to this may be added twenty drops of tincture of aconite, two drachms of chloroform, or two ounces of sulphuric ether. Hot water should be applied to the abdomen by means of rugs, and the white liniment or a mustard plaster may also be applied to this region before the hot water is resorted to.

When the horse can eat, the diet should consist of linseed and bran mashes, and no hay or other solid food ought to be given for some days.

Worms.

Worms are often troublesome to horses, by causing irritation of the intestines, and unthriftiness and debility. There are several kinds of worms which we need not, for lack of space, describe, especially as the treatment is nearly the same for all. This generally consists in the administration of a purgative, followed by an ounce dose of oil of turpentine in flour gruel, or well mixed in a pint of milk; or one or two one-drachm doses of tartar emetic in a little mash, followed by half a dozen one-drachm doses of powdered sulphate of iron—one dose morning and evening.

LAMENESSES.

The horse is, from the nature of his work, much exposed to lameness, and this very often becomes permanent, and more or less reduces his value. Lameness may be due to many causes, and these may be in operation in any part of the limb or limbs; sometimes injury or disease of other parts of the body will also produce lameness. We will notice some of the more common forms of lameness, with their causes and treatment.

Sprains.

Sprains may occur to tendons and ligaments, less frequently to muscles, and this injury may be more or less severe, and cause a proportionate degree of lameness. Ligaments and tendons, as well as muscles, during violent efforts or from twists, may be over-stretched and their fibres torn, or the injury to them may be brought about gradually, as in some tendons and ligaments of the lower part of the limbs. No matter where sprains occur, more or less prolonged rest, as complete as possible, is essential to rapid and permanent recovery. Next to rest comes reparative treatment, and this will vary somewhat according to the seat and the nature of the sprain. When it is quite recent, attempts must be made to check the swelling and inflammation that ensues, and with this object in view, the application of water—cold or hot—or soothing and evaporating lotions, is resorted to. All are beneficial according to the assiduity with which they are applied. The water should either be always rather cold or as hot as the horse can bear it. When it can be done, the part should be enveloped in bandages or swabs, so as to retain and distribute the moisture or lotion. Perhaps the best lotion is that composed of Goulard's extract, subacetate of lead and spirit in equal parts, with eight to ten parts of water. When the pain and swelling have subsided somewhat, then a mild stimulant may be applied, such as the acetic acid liniment already alluded to. Gentle exercise may also be allowed if there is no lameness, and continued until the horse is fit for work.

Sprain of the Back Tendons.

This is perhaps the most frequent sprain to which heavy horses are liable, and it may occur either in the fore or hind legs. There is swelling, heat, and pain on pressure of the injured part, and lameness corresponding to the extent of the injury. A shoe raised two or three inches at the heels, should be put on the foot of the sprained leg, and the general treatment prescribed above resorted to. If the injury is very severe and considerable thickening remain, it may be advisable to apply the biniodide of mercury ointment, or cantharides ointment to it; it may even be necessary to "fire" the part in order to effect efficient recovery. Instead of this, the projection of cold water from a hose for from ten to twenty minutes, three or four times a day, may be advantageous in expediting a cure; indeed, this may be carried out from the very commencement, the lead lotion being applied in the intervals. Sprains of these or other tendons or ligaments in this region may also be treated after the method recommended by Captain Hayes, which consists in enveloping the part in cotton wool, and bandaging tightly in such a manner as to ensure uniform pressure. This bandaging may be employed after applying the hot or cold water or lotion, and is most conveniently carried out as he directs. "Take about half-a-pound of cotton wool, and a cotton bandage (such as can be got in any chemist's shop) about three inches broad and six yards long. First of all, wrap loosely round the leg a piece of soft cotton cloth, or put on an ordinary flannel bandage, as the contact of wool sometimes causes irritation to the skin. Place a little cotton-wool at each side of the leg at the place where it is desired to commence, and loosely wrap the bandage over it, adding at each turn more cotton wool, some of which should also be placed at the front and back of the leg, until there is a layer about four inches thick round the part. As the bandage is passed around the leg it may be gradually tightened until at last it is made very tight, when it can then be secured by sewing or by tapes. The bandage should be removed after twenty-four hours, the part rubbed firmly upwards by the hand (the leg being held up during this massage, and flexed and extended); and a fresh bandage of the same kind put on. The bandage may then be removed morning and evening, and the part hand-rubbed and passively worked by bending the joints without causing the horse to move." The tendon may be rubbed with the stimulating liniment during the massage; if the hair is long it may be clipped off. The cotton wool should be of the ordinary kind—soft and elastic, and it is better to have it fresh at each application. The diet should be rather laxative, and green forage ought to be given if it can be procured.

The high-heeled shoe should not be kept on the foot for more than a fortnight, when its heels may be gradually lowered. If considerable improvement has not taken place in three weeks of this treatment, a charge may be applied to the tendon. This is variously composed, but the usual ingredients are Burgundy pitch and bees'-wax, four parts of each; when these are melted in an iron ladle two parts of mercurial ointment are stirred in. When moderately warm this is plastered in a thick layer over the leg by means of a spatula or hard brush, pieces of cotton-wool being stuck on the skin and the hollows on each side of the tendon as the smearing goes on. Over these the mixture is to be daubed, and when sufficient has been applied to make the leg a rounded mass, a long cotton bandage is tightly bound over it, the mixture being laid upon this at every turn and cotton wool placed between each layer, so as to effect equable and firm pressure. If at any time the layers should become loose they may be plastered with the warmed mixture. From three to five weeks is sufficiently long to keep on this bandage.

Splints.

Splints are bony tumours which form either inside or outside the leg—usually the former—and generally in the neighbourhood of the small splint bones. They most frequently form in young horses, and are most readily seen when the limb is looked at in front. It is usually when they are forming that they cause lameness, but when they are so situated as to interfere with movement, the lameness may be permanent. There is heat, and pain on manipulation.

The best treatment for the amateur to adopt consists in the application of Goulard's lotion already described, this being poured on to a woollen or cotton bandage enveloping the leg where the splint is forming. After a few days of this treatment a little piece of the biniodide of mercury may be rubbed into the skin over the tumour. Exercise should be allowed on soft ground.

Ringbone.

This is a deposit of bony matter on the surface—front or sides—of the pastern bones, and is generally very serious, owing to the deposit interfering with the tendons and ligaments covering it. It is most frequently observed on the front pasterns.

The treatment should be the same as for splints, but it must be long continued and the horse should be rested as much as possible, the stall or loose box being laid with peat moss litter. In chronic cases firing may be necessary.

Side-bone.

This name is given to the plate of elastic cartilage on each side of the foot towards the heels when it becomes more or less rigid from the deposition of bony matter in its substance. This loss of elasticity generally occasions lameness, which is most marked and serious when both plates—inside as well as outside—are involved. Side-bone is most frequently seen in heavy draught horses, and especially those with coarse hairy legs. There is no doubt an hereditary predisposition to this grave alteration, but it is not improbable that its occurrence is often due to faulty management of the horse's foot in the process of shoeing, one side of the hoof being left higher than the other by the shoer; and so the foot and limb are twisted or bent outwardly or inwardly when the horse puts his weight on the leg. It is most essential to the healthy condition and continued integrity of these cartilages, that the hoof be at all times properly levelled when the horse is being shod. The existence of side-bone is readily ascertained by pressing firmly on the cartilage with the thumb. If it is altered in texture it feels hard and rigid. The horse generally goes more or less lame, though instances occur in which the gait is scarcely, if at all, altered. The fore-feet are those specially liable to side-bones, and horses in towns appear to have them far more frequently than those in the country. External injury—such as a tread—may give rise to side-bone by setting up inflammation in the cartilage, the whole or only a portion of which may be implicated in the change.

Treatment.—With regard to treatment, it must be admitted that it is most difficult to stop the progress of the change in the cartilage when it has once commenced; much more difficult is it to restore the altered cartilage to its normal condition. Poultices and blisters do not produce much result, and all other kinds of treatment hitherto proposed have proved of little avail. At the very commencement, if it could be ascertained that side-bone was forming, long continued rest and cold applications—such as cold water—might check the change, but it is very difficult to ascertain positively when this is beginning. Cold water at first, and blisters afterwards may be tried. The foot should be carefully shortened and levelled, and the frog brought to bear largely on the ground. If the horse must be worked, then a bar shoe resting to a great extent on the frog, or an ordinary shoe slightly curved upwards at the toe and heel, must be put on.

Bone Spavin.

This is usually a bony enlargement inside, and at the lower part of the hock; in some instances there is little, if any, enlargement, but two or more of the bones of that joint may be joined together, and there may also be ulceration between them. There is more or less stiffness or lameness, according to the extent and seat of the disease. The horse rests the leg very much, and goes on the toe of the foot. When he first begins to move, the lameness is much greater than it is after travelling for some time. The lameness is sometimes very perceptible when the horse is moved in the stall.

Treatment.— To be at all beneficial, treatment must be undertaken early. Absolute rest is indicated, and if the horse could be rendered immovable in the affected joint, there would be a good chance of stopping the progress of spavin. But this is not possible, and all that can be done is to keep the horse quiet, a stall being preferable to a loose box, and the animal can be kept tied up for some time. To ease the front of the joint, a high-heeled shoe should be placed upon the foot, and warm or cold fomentations applied to the hock for some time. Then biniodide of mercury ointment should be rubbed into the skin over the spavin at intervals of a week or so. This treatment ought to be continued for six weeks or two months, when the result should be tested. If the lameness has not disappeared, then firing should be resorted to, points being employed instead of lines.

Thoroughpin.

Thoroughpin is the name given to distention of the sheath of the tendon of the hind foot at the upper and back part of the hock. The tendon may be strained or its sheath injured at this point, and the swelling may be pushed from one side to the other, hence the name. This condition is much more frequent in heavy than light horses, and in those with short hocks. There may or may not be lameness, but the swelling is unsightly.

Treatment.—If the sprain is recent, then rest is indicated, and the application of a high-heeled shoe to the foot of the affected leg. Fomentations with warm water may be resorted to for some days, after which tincture of iodine may be painted over the swelling every day until the skin becomes slightly blistered, or the biniodide of mercury ointment may be applied twice or three times at intervals of a week. When the swelling is chronic, then the spring truss made by veterinary instrument makers for effecting pressure on this part of the hock should be tried.

Bog Spavin.

Bog spavin is a soft swelling on the front and inner part of the hock, above the seat of bone spavin, and is due to distension of the capsule of the joint. When the distension is great there is also swelling in the seat of thoroughpin, from the capsule being pushed upwards and backwards. Bog spavin may appear without any assignable cause, but there is generally a sprain or series of sprains of the hock, to which it owes its production. Horses which start great loads, and heavy stallions whose hocks are severely strained in covering, are those which most frequently show bog spavin and thoroughpin, and especially if their hocks are short.

Treatment.—This should be the same as for thoroughpin, the employment of the spring truss being even more beneficial for this condition than for the one just mentioned.

Wind Galls.

Wind galls are merely distentions of the sheaths of tendons below the knees and hocks, due either to rheumatism, sprain, or hard work. They may or may not be accompanied by lameness; if they are, then the soft puffy swelling is hot, and painful on pressure. They are most frequently seen about and immediately above the fetlocks, especially those of the hind limbs.

Treatment.—If there is lameness, then the treatment should be as for sprain of the tendons. If there is no lameness, but merely distension, then equable pressure by means of bandages is the simplest and readiest treatment. When rest can be allowed for some time, a charge (already described) may be applied to the leg, where the swelling is.

Lymphangitis or Weed.

This is an inflammation of the lymphatic vessels of the hind legs, usually only one, to which heavy horses are more especially liable. It is due to derangement of digestion or over-feeding when not working, and appears most frequently on Monday morning, after Sunday's rest.

Symptoms.—There is much fever and great pain in the affected leg, which is swollen from the foot to above the hock, and the horse moves it with hesitation and difficulty. After a time the inflammation subsides, but it is always likely to recur after the first attack, leaving a gradually increasing thickness of the leg until at last it sometimes is greatly developed in size.

Treatment.—In the acute stage, a strong dose of physic should be administered, and hot water fomentations applied to the leg, from the stifle to the foot, continuously for hours. After this the leg should be well dried, then rubbed with the ammonia and turpentine liniment, and bandaged with flannel. The diet should be sloppy mashes and a small quantity of good meadow hay for a few days. When the inflammation has subsided, then the leg should be well hand-rubbed upwards, and frequently during the day. Sugar of lead lotion may also be sponged over it once a day. To prevent recurrence of the inflammation, care should be taken to reduce the amount of rich food (oats and beans) given on resting days, and increase the allowance of hay.

Grease.

Grease is an inflammation of the skin (sebaceous follicles) of the lower and back part of the legs, generally the hind ones, which gives rise to a thin, greasy, and peculiar smelling discharge, accompanied by a certain degree of soreness and stiffness. It is most frequently seen among coarse, hairy-legged horses which are badly attended to, and is somewhat common in damp and dirty stables. The exciting cause is generally cold and wet acting upon a dirty skin, and the prevention is cleanliness and drying the legs when the horse returns to the stable.

Treatment.—If the skin is much inflamed and sore, and discharging the offensive secretion, it may be necessary to clip away the long hair, and foment and poultice the part. After a day or two of this treatment, an astringent lotion should be applied. A very good lotion is composed of two parts of lead acetate and one and a-half part zinc sulphate dissolved in about thirty parts of water. This should be well shaken up when about to be used, when it forms a white lotion which should be applied to the diseased skin by means of a piece of sponge.

Should it be necessary to work the horse before the skin is perfectly healthy, and there is a likelihood of the limbs becoming wet and dirty, it is advisable to cover the heels and back of the legs with a mixture composed of white lead one part, linseed oil two parts. This should be applied by means of a brush.

Thrush and Canker.

These are diseases commencing in the frog of the hoof, "thrush" being generally the prelude to the serious condition named "canker." Thrush is due to several causes, the chief of which are paring the frog when the horse is being shod, and not allowing it to come in contact with the ground. There is a foul-smelling discharge from the cleft of the frog, which becomes ragged and wasted, and if the exciting cause is allowed to go on and no treatment is adopted, then the horn becomes underrun by the matter, and separated from the living parts beneath; these throw out fungous cauliflower-like growths, which may extend over the sole, and even invade the wall of the hoof. This constitutes canker, which is usually seen in coarse-bred heavy horses, and is generally due to neglect; it may appear in fore and hind feet, but perhaps the latter are most frequently attacked.

Treatment.—Cleanliness in the stable, and removing dirt and stones from the hoofs when the horse returns from work, will tend to prevent the occurrence of thrush, provided the shoeing-smith does not mutilate the frog with his knife, and that part is allowed to meet the ground. When a discharge appears from the cleft of the frog, this part should be well cleaned out to the very bottom by means of pledgets of tow; then a piece of tow dipped in Stockholm tar should be pushed to the bottom of the cleft and left there for a day or two, when it ought to be removed, and replaced by a similar dressing until all discharge has ceased and the cleft has filled up to its natural state. Sometimes calomel passed into the cleft in the same way answers the purpose. If the frog is generally unsound, all the loose parts should be cut away and the tar smeared over the surface. Getting the frog to meet the ground is always most beneficial.

For the cure of canker skilled assistance is necessary, as an operation is usually required to fully expose the diseased surface. Then caustics and astringent powders are needed to destroy the fungous growths, while pressure and dryness are adjuncts which must not be overlooked.

Injuries to the Foot.

The foot is more exposed than any other part of the body to injuries of various kinds, such as treads, contusions, wounds from sharp objects while travelling on the road, pricks and bruises in shoeing, splitting of the hoof (sand crack), bruise of the sole (corn), &c.

When the injury occurs to a part enclosed in the hoof, it is generally necessary to relieve the sensitive parts from pressure by removing the horn from over and around it, and preventing the shoe from touching it. When the inflammation runs high and there is much pain, fomentations with hot water and poulticing are necessary, but these must not be continued very long, and, as a general rule, they should be succeeded by dry dressings. For injuries in which the hoof is concerned, after the inflammation has been subdued, Stockholm tar is an excellent dressing.

Wounds.

Wounds are of different kinds, according to their mode of production, such as incised, punctured, contused, &c. The incised wound is that which is generally most readily repaired. When there is hæmorrhage it should be checked as soon as possible, and this can sometimes be effected by the application of cold or hot water, bandaging up the wound, applying pressure, or tying the bleeding vessel or vessels. Some chemical agents, such as perchloride of iron, are sometimes employed to check bleeding.

If the wound is not large and the part can be bandaged, then its edges should be brought together and the bandage applied, a piece of lint or tow being previously placed upon the wound. If it can be done, it is often beneficial to bring the edges of the wound into apposition by means of one or more stitches; or by brass pins passed through the skin, and a piece of tow or twine wound in figure of 8 fashion around the heads and points.

Bleeding from a punctured wound can generally be stopped by plugging it firmly with tow, lint, or any similar substance. The air should be excluded as early and as perfectly as possible from all wounds, so that after dirt and any other extraneous matters which may have gained access to them are removed, they may be carefully protected by tincture of myrrh, powdered boric acid, iodoform, or other antiseptic agent.

When the wounds are large and contused, it is not generally

advisable, nor is it often possible, to close them by sutures or close bandaging, as the dead portions have to be removed by the natural process of sloughing or suppuration. This can often be expedited by fomentations with warm water.

Broken knees are more frequent in fast moving light horses than in heavy horses, though the latter sometimes have these joints badly damaged by falling on them. When such an accident takes place, the wound should be freed from dirt and grit by gentle washing with a sponge; a piece of lint ought then to be placed over the injury, and maintained there by a bandage. When the wound is not deep or very contused, some Canada balsam, spread on a piece of lint and laid upon the wound after it has been cleaned and dried, often has an excellent effect, being allowed to remain until the wound has healed. When the wound is deep and contused, and the joint probably opened, then the legs should be kept immovable by a splint and a starch bandage extending from above the knee to the foot, the portion of the bandage covering the wound being cut out after it has dried, in order to permit the injury to be dressed. This dressing should consist of boric acid or iodoform powdered over the wound. The horse should not be allowed to lie down, and it is generally advisable to have him slung, to prevent his falling, until the wound is healed.

Mange.

This is a rare disease among heavy horses in well managed stables where grooming is carefully carried out. When introduced, however, among horses which are not well looked after, it is very troublesome, and often damaging to them. It is caused by microscopic insects, one kind of which causes mange of the body, and is the most annoying to the horse; another kind infests the neck at the root of the mane; and a third kind locates itself on the thickly-haired legs of heavy horses. They all cause intense itching, which compels the animals to bite and rub themselves almost continuously. The insect that infests the body also produces shedding of the hair in patches, and raw places and crusts on the skin.

Mange is very contagious, and the parasites can pass directly from affected to healthy horses, as well as by means of harness, clothing, straw, &c.

Treatment.—Cleanliness is a potent barrier to the extension of mange. Affected horses should be well washed with warm water and soft soap, applied by means of a scrubbing-brush, then when dry the skin must be dressed with some agent that will kill the parasites. For the mane-and-tail insect, and also for that inhabiting the legs, one dressing of an ointment composed of one part of tar oil and six parts of palm oil, will generally suffice, the ointment being washed off in two or three days. For body mange, it is most advantageous, after washing the skin with soap and water, to soak it for some hours with a solution of carbonate of potass and oil before this ointment is applied, and it is usually necessary to repeat the treatment.

In addition to treating the animals, it is essential that clothing, harness, stable fittings, and everything else with which affected horses may have been in contact, should be cleansed and dressed with a solution of carbolic acid, one part to five or ten of water.

RINGWORM.

Ringworm is due to the presence of a microscopic vegetable parasite, which grows on the skin in such a manner as to produce more or less circular bare patches covered by a thin crust. It does not cause so much itching as mange, though there is some; but it renders the skin unsightly, and may cause it considerable damage if it is allowed to exist for a considerable time. It is oftenest seen perhaps in heavy horses, and more particularly those which are young. It is very contagious.

Treatment.—This may be the same as that recommended for mange, but the treatment may be limited to the affected parts and a little distance beyond them.

SHOEING.

The management of horses' feet, with the object of keeping them healthy, is perhaps not so important with heavy as light horses; nevertheless, it is necessary that it should not be overlooked in attending to the welfare of the former, and therefore we may briefly allude to the following rules, the observance of which will be beneficial to horse-owners:—

(1) Heavy horses should be shod, or the old shoes be removed, at least once a month.

(2) Then the hoofs should be reduced to a proper length, and evenly levelled, so that one side of the foot will not be higher than the other.

(3) The frog and sole should not be pared, interference with them being limited to removal of any loose portions.

(4) The shoe should not be heavier than is necessary to withstand wear for a certain period—say a month.

(5) It should be made to fit the hoof—that is, the full size of the latter; and it ought to be level on the surface on which the hoof rests.

(6) It should be attached to the hoof with as few and as small nails as may be necessary to keep it securely on the hoof.

(7) The nails should not be driven higher into the hoof than is required to obtain a sound and firm hold.

(8) When the shoe is nailed on and the clenches laid down, the front of the wall should not be rasped at all, but left with its natural polish and in all its strength.

(9) If possible, the frog should be allowed to come in contact with the ground.

INDEX.

Action of Clydesdales, 110
Action of Shire Horses, 20
Admiral, 26
Ayrshire Clydesdales, 91

Bald Horse 93, 1778, 11
Bar None 2388, 28, 137
Bitting and Breaking, 170
Blake's Farmer, 44
Blaze 183, foaled 1770, 11
Blundeville, Sir Thomas, on English Horses, 4
Bog Spavin, 208
Bone of Shire Horses, 20
Bone Spavin, 206
Brady's Briton, 62
Breaking, 170
Breeding Heavy Cart Horses for Street Work, 122
Breeds :—
  The Clydesdale Horse, 75
  The Shire Horse, 1
  The Suffolk Horse, 37
Bronchitis, 193
Brood Mares, 142
Broomfield Champion, 81
Bury Victor Chief 11105, 34

Cæsar on English War Horses, 2
Canker, 210
Capon's Duke, 65

Cart Horses at Smithfield Fair in 1154, 4
Cart Horses on the East Coast, 10
Castlereagh, 131
Catarrh, 186
Catlin's Duke, 66
Champion 419 (Styche's), 12
Champion Horses at London Shows, 26
Characteristics of Clydesdales, 117
Characteristics of Shire Horses, 20
Characteristics of Suffolk Horses, 72
Chart showing descent of Suffolk Horses (*to face*) 59
Clyde *alias* Glancer, 83
Clydesdale and Shire Cross, 135
Clydesdale Horse, The, 75
Clydesdales in Cumberland and Aberdeen, 94
Colic, 199
Colour of Suffolk Horses, 50
Colours of Early Shires, 11
Colts, Two-Year-Old, Management of, 172
Congestion of the Lungs, 195
Crisp of Ufford's Horses, 41, 43
Cupbearer, 51, 69
Cullum, Sir T., on Suffolk Horses, 40

Darnley, 102

Demand for Heavy Horses, 132
Diseases and Injuries of Heavy Horses, 183
Dodman, English Cart Horse (1780), 9
Draught, 174
Drew's, Mr., Purchase of English Mares, 124
Drumore Stud, 87

Early Clydesdale Sires, 80
Early History of Shire Horse, 2
Early Records of Suffolks, 37
Early Stud Book, Records of Shires, 6
Edward's Old Briton, 59
English Horses in Scotland, 123
Enterprise of Cannock, 29

Farm Management of Heavy Horses, 159
Farmer, 112
Feeding and Watering, 177
Feeding Farm Horses, 173
Feeding London Work Horses, 156
Feeding Town Horses, 180
Fever, 184
Flemish Horses at Holkham, 39
Flemish Stallions in Lanarkshire, 75
Foaling, 162
Foals, Management of, 163
Fulton's Ruptured Horse, 81

G. 890, Shire Stallion, 11
Galloway Clydesdales, 86
Geldings, 149
Gilbey, Sir Walter, on Old English War Horse, 2
Glancer, Clydesdale Stallion, 81
Glanders and Farcy, 191

Grease, 209
Grooming, 178

Hair on Shire Horses, 9
Harnessing and Yoking, 82
Harold 3703, 31
Henry VIII. and Horse Breeding, 4
Hereditary Diseases, 138
Hitchin Conqueror, 4458, 33
Hours of Work of Farm Horses, 174
Housing Farm Horses, 175

Importations from the Continent, 8
In Foal Mares, Management of, 16
Inflammation of the Bowels, 200
Inflammation of the Lungs, 196
Influenza, 190
Injuries of Heavy Horses, 183
Injuries to the Foot, 211
Introductory Essay to Clydesdale Stud Book, 77

Judicious Mating, 143

Keir Peggy, 103
Kintyre Clydesdales, 92

Lamenesses, 201
Laminitis, 198
Lampit's Mare, 80
Leading Clydesdale Tribes, 102
Lincolnshire Lad 13, 24
Lincolnshire Lad Mares, 124
Littering Horses, 176
Logan's Twin, 104
London Work Horses, 145
Lord Ailsa, 133

Lord Erskine, 109
Lord Lyon, 112
Lymphangitis and Wind, 209

Management of Heavy Horses, 159
Management of Mare and Foal, 163
Mange, 213
Manseller (Oldacre's) 11
Mares, Management of, 161
Marston sold for 500 guineas, 12
Matchless 1609, 14
Modern Clydesdales and their Characteristics, 101
Modern Shire Sires, 23

Old English War Horse, 2
Old Times, Clydesdale Horse, 113
Orton, Dr., Views on Physiology, 135

Packington Blind Horse, 11
Paterson of Lochlyoch, 77
Pedigree, 143
Pedigree Influence and Traditions, 11
Physiology of Breeding, 134
Pleurisy, 197
Points of Clydesdales, 117
Points of Shire Horses, 20
Points of Suffolk Horses, 72
Prince Lawrence, 129
Prince of Avondale, 128
Prince of Wales, 106
Prince William 3956, 30
Purchase of English Horses by Scotch Dealers, 123

Queen Anne's State Equipages 6

Rations of London Work Horses, 156
Rearing Horses, 144
Reynolds, Mr., History of English Cart Horse, 8
Rheumatism, 198
Ringbone, 205
Ringworm, 214
Rokeby Harold, 35
Royal Albert, 1885, 16
Ruler (1773), 11

Scotch Horses in England, 125
Selection of Stallions, 141
Shire and Clydesdale Cross, 133
Shire Horse, 1
Shire Horse Show, 21
Shire Horse Stud Book, 6
Shoeing, 179, 215
Side-bone, 205
Sinclair, Sir John, on Scotch Horses, 79
Soundness in Suffolks, 54
Spark 2497, 26
Spavin, 206, 208
Splints, 204
Spooner, W. C., on Cross Breeding, 135
Sprain of the Back Tendon, 202
Sprains, 202
Stables, 175
Stallions, Selection of, 140
Statistical Account of Scotland, 79
Staunton Hero 3913, 31
Stradbroke, Lord, on Suffolk Horses, 53
Strangles, 187
Suffolk Horse, 37
Suffolk Stud Book, 38
Suffolks as Agricultural Horses, 55
Suffolks in London Streets, 56

Sweet William, sold in 1778 for 350 guineas, 12

Three-Year-old Fillies and Geldings, 172
Thoroughpin, 207
Thrush and Canker, 210
Topsman, 115
Two-Year-olds, Management of, 170
Typical Shire Brood Mare, 20

Uniformity of Suffolk Horses, 41

Veterinary Inspection, 54
Vulcan 4145, 32

Watering Horses, 177
Weaning, 165
Wedgewood, 64
Weight in Shire Horses, 21
What's Wanted 2332, 14, 137
Whitefaced Boxer, 68
William the Conqueror 2343, 13
Windgalls, 208
Worms, 201
Wounds, 212
Wright's Farmer's Glory, 45

Yearlings, Management of, 168
Young, Arthur, on Suffolk Horses, 40

ONLY ONE ADDRESS.

Day & Sons, Crewe.
Established 1840.

HORSE, CATTLE, SHEEP & DOG MEDICINES.

Largest Veterinary Providers in the World.

"Breeders and Owners of Stock can rely on the Preparations supplied by Messrs DAY & SONS, of Crewe."—*Live Stock Journal.*

DAYS' "BLACK DRINK"

(Often called the "Magic Drink") cures like a charm Colic or Gripes and Chills in Horses and Cattle; instantly relieves Hoven or Blown Cattle and Sheep; stops Scouring, and is the best general Stimulant and Tonic for Calves and Lambs.

Matchless as a Restorative and Painkiller after Lambing and Calving; for Fatigue in Hunters and overworked Horses, and in all cases where nature flags.

Price 5/- per Quarter Dozen, post paid, or 19/- per Dozen Bottles in Boxes. Carriage paid.

DAYS' RED DRINK,
Or Cow Drench.

For Costiveness, Loss of Cud, Garget, Colds, Fever, Hide-bound, &c. Prevents Milk Fever, and Cures Bad Cleansing. Prices—1/- per packet; for Ewes, 3/6 per dozen.

ARRANGED FOR **Disorders of Horses, Cattle, and Sheep.**	ARRANGED FOR **Disorders of Horses and Colts.**
Prices: £5, £2 4s. and £1 4s. (with Guide, "Everyday Farriery"). Carriage Paid.	Prices: £5, £2 14s. and £1 4s. (with Guide, "Everyday Farriery"). Carriage Paid.

Special Lambing and Calving Cases: £1 1s., £2 2s., £3 3s. and upwards, carriage paid.

Note that we have only "One" Address, DAY & SONS, CREWE.

Advertisements.

Flockmasters secure better Breeding results and reduced mortality amongst lambs

BY USING

J. RANDS & JECKELL'S

SHEEP AND LAMB

SHELTERING CLOTHS.

SHEEP AND LAMB
SHELTERING CLOTHS.

"Indispensable to every Flockmaster."

20 yards long, 3 feet deep, with Brass Eyelets and Cords for fixing to Hurdles.

6d., 9d. & 1s. per yard.
Rot Proof 1/6 per yard.

THE "DUPLA"
SHELTERING CLOTHS.

"The Flockmaster's Friend."

20 yards long, 6 feet deep, with Brass Eyelets and Cords for fixing to hurdles.

1s., 1/6 & 2s. per yard.
Rot Proof 3s. per yard.

Carriage Paid on Orders above £2; 5 per cent. Discount for Cash.

WRITE IMMEDIATELY FOR NEW PAMPHLET ON

"SHELTER FOR SHEEP AND LAMBS."

Post Free on application to

J. RANDS & JECKELL, IPSWICH,

Sack, Rick-Cloth and Tent Manufacturers by Special Royal Warrant to H.R.H. The Prince of Wales.

Advertisements.

LIVE STOCK JOURNAL.

Friday. (Illustrated.) Fourpence. Established 1874.

The only Paper in the United Kingdom wholly devoted to the Interests of Breeders and Owners of all varieties of Live Stock. Contains contributions from the highest authorities on all matters relating to the Breeding, Feeding, and Veterinary treatment of Domesticated Animals, and Illustrations of the more celebrated specimens.

Gives the fullest and earliest reports of Agricultural Shows, Stock Sales, Sheep Sales, and Lettings, whilst its Herd and Flock Notes and Notes from the Stables contain much valuable and interesting information. Prominence is given in the columns of the JOURNAL to correspondence on all questions of interest to Country Gentlemen, Breeders, and Exhibitors.

Subscription:—3 Months, post free, 5s.; 12 Months, 19s. 6d.; Foreign Subscription, £1 2s. per year.

AGRICULTURAL GAZETTE.

A WEEKLY JOURNAL OF FARMING & MARKET GARDENING.

MONDAY. (ILLUSTRATED.) TWOPENCE.
Established 1844.

Has for many years stood at the head of the English Agricultural Press. Unequalled as a comprehensive practical paper. All branches of farming—crops, live stock and dairy—are fully discussed by leading practical authorities. Market intelligence and reviews of the grain and cattle trades are special features. Prompt replies given to questions in all departments of farming. Veterinary queries answered by a qualified practitioner. The Market Gardening section deals fully with the production and Marketing of vegetables and fruit. Special Articles on Cultivation, Manuring, New Varieties, &c., appear weekly.

Subscription:—3 Months, post free, 2s. 9d.; 12 Months, 10s. 10d.

BAILY'S MAGAZINE
OF
SPORTS & PASTIMES.

Racing, Hunting, Shooting, Yachting, Rowing, Fishing, Cricket, Football, &c.

This well-known monthly contains articles written by the best authorities on every phase of British Sport; and in addition to the usual Frontispiece—a Steel Plate Portrait of an eminent sportsman—other Illustrations of well chosen subjects and of the highest artistic merit are given.

Of all Booksellers and at all Bookstalls, 1s. Or by post direct from the Office, 14s. per year.

India Proofs of any of the Engraved Portraits of which some 800 have appeared, 2s. 6d. each.

VINTON & CO., LTD., 9, New Bridge Street, London, E.C.

Advertisements.

LIVE STOCK HANDBOOKS.

No. 1.
SHEEP: Breeds and Management.
SECOND EDITION.

By JOHN WRIGHTSON, M.R.A.C., F.C.S., President of the College of Agriculture, Downton; Professor of Agriculture in the Royal College, London, &c. 236 pages, demy 8vo, cloth, gilt lettered, with 24 full-page Illustrations of the various Breeds. 3s. 6d.; post free, 3s. 10d.

No. 2.
LIGHT HORSES: Breeds and Management.
SECOND EDITION.

By W. C. A. BLEW, M.A.; WILLIAM SCARTH DIXON; Dr. GEORGE FLEMING, C.B., F.R.C.V.S.; VERO SHAW, B.A., &c. 226 pages, demy 8vo, cloth, gilt lettered, 28 full-page Wood Engravings of the various Breeds. 3s. 6d.; post free, 3s. 10d.

No. 3.
HEAVY HORSES: Breeds and Management.
SECOND EDITION.

By HERMAN BIDDELL; C. I. DOUGLAS; THOMAS DYKES; Dr. GEORGE FLEMING, C.B., F.R.C.V.S; ARCHIBALD MACNEILAGE; GILBERT MURRAY; and W. R. TROTTER. 224 pages, demy 8vo, cloth, gilt lettered, with 29 full-page Illustrations.

No. 4.
CATTLE: Breeds and Management.

By WILLIAM HOUSMAN and Professor J. WORTLEY AXE. 272 pages, with 34 full page Illustrations.

No. 5.
PIGS: Breeds and Management.

By SANDERS SPENCER and Professor J. WORTLEY AXE. 180 pages with 20 Illustrations.

3s. 6d. each, or post free 3s. 10d.; or the set of 5 volumes, if ordered together direct from the office, 17s. 6d., carriage free.

VINTON & CO., LTD., 9, New Bridge Street, London, E.C.

MORTON'S
HANDBOOKS OF THE FARM.

The aim of the Series is to display the means best calculated to secure an intelligent development of the resources of our soil, and with the assistance which advanced Chemical investigation provides, to direct those engaged in Agricultural Industry towards the most successful results. Each Book is complete in itself, and the short Series of handy volumes, by various writers, who have been specially selected, forms a complete HANDBOOK OF THE FARM, which is abreast of the enterprising man's every-day requirements, and enables him economically to utilise the advantages which an everwidening science places within his reach.

PRICE 2s. 6d. EACH.

No. I. **CHEMISTRY OF THE FARM.**
By R. WARINGTON, F.R.S.
Revised and Enlarged. Eleventh Edition.

No. II. **LIVE STOCK.**
By W. T. CARRINGTON, G. GILBERT, J. C. MORTON, GILBERT MURRAY, SANDERS SPENCER, and J. WORTLEY-AXE.

No. III. **THE CROPS.**
By T. BORWICK, J. BUCKMAN, W. T. CARRINGTON, J. C. MORTON, G. MURRAY, J. SCOTT, and R. HENRY REW.

No. IV. **THE SOIL.**
By Professor SCOTT and J. C. MORTON.

No. V. **PLANT LIFE.**
By MAXWELL T. MASTERS, F.R.S.

No. VI. **EQUIPMENT.**
By WM. BURNESS, J. C. MORTON, and GILBERT MURRAY.

No. VII. **THE DAIRY.**
By JAMES LONG and J. C. MORTON.
Revised and Enlarged.

No. VIII. **ANIMAL LIFE**
By Professor BROWN, C.B.

No. IX. **LABOUR.**
By J. C. MORTON.

No. X. **WORKMAN'S TECHNICAL INSTRUCTOR.**
By WALTER J. MALDEN.
Illustrated.

In crown 8vo volumes, the complete set of ten volumes, if ordered direct from the Office, carriage free for £1 2s. 6d.

VINTON & CO., LTD., 9, New Bridge Street, London, E.C.

www.ingramcontent.com/pod-product-compliance
Lightning Source LLC
Chambersburg PA
CBHW030403250426
43670CB00050B/434